Biomimetics - Bridging the Gap

Edited by Ziyad S. Haidar

Published in London, United Kingdom

Biomimetics - Bridging the Gap
http://dx.doi.org/10.5772/intechopen.100667
Edited by Ziyad S. Haidar

Contributors
Ching-Cheng Huang, Masashi Shiotsuki, Nikolaos K. Uzunoglu, Halima Feki Ghorbel, Awatef Guidara, Yoan Danlos, Jamel Bouaziz, Christian Coddet, Ziyad S. Haidar, Pugalanthipandian Sankaralingam, Poornimadevi Sakthivel, Vijayakumar Chinnaswamy Thangavel, Emad Kamil Hussein, Kussay Ahmed Subhi, Batool Mardan Faisal, Thiago Santos, Caroliny Santos, Samir Ghouali, Muhammad Asyraf

First published in London, United Kingdom, 2023 by IntechOpen
IntechOpen is the global imprint of INTECHOPEN LIMITED, registered in England and Wales, registration number: 11086078, 5 Princes Gate Court, London, SW7 2QJ, United Kingdom

British Library Cataloguing-in-Publication Data
A catalogue record for this book is available from the British Library

Additional hard and PDF copies can be obtained from orders@intechopen.com

Biomimetics - Bridging the Gap
Edited by Ziyad S. Haidar
p. cm.

This title is part of the Biomedical Engineering Book Series, Volume 17
Topic: Bioinspired Technology and Biomechanics
Series Editor: Robert Koprowski
Topic Editor: Adriano Andrade

Print ISBN 978-1-80356-896-6
Online ISBN 978-1-80356-897-3
eBook (PDF) ISBN 978-1-80356-898-0
ISSN 2631-5343

IntechOpen Book Series
Biomedical Engineering
Volume 17

Aims and Scope of the Series

Biomedical Engineering is one of the fastest-growing interdisciplinary branches of science and industry. The combination of electronics and computer science with biology and medicine has improved patient diagnosis, reduced rehabilitation time, and helped to facilitate a better quality of life. Nowadays, all medical imaging devices, medical instruments, or new laboratory techniques result from the cooperation of specialists in various fields. The series of Biomedical Engineering books covers such areas of knowledge as chemistry, physics, electronics, medicine, and biology. This series is intended for doctors, engineers, and scientists involved in biomedical engineering or those wanting to start working in this field.

Meet the Series Editor

Robert Koprowski, MD (1997), Ph.D. (2003), Habilitation (2015), is an employee of the University of Silesia, Poland, Institute of Computer Science, Department of Biomedical Computer Systems. For 20 years, he has studied the analysis and processing of biomedical images, emphasizing the full automation of measurement for a large inter-individual variability of patients. Dr. Koprowski has authored more than a hundred research papers with dozens in impact factor (IF) journals and has authored or co-authored six books. Additionally, he is the author of several national and international patents in the field of biomedical devices and imaging. Since 2011, he has been a reviewer of grants and projects (including EU projects) in biomedical engineering.

Meet the Volume Editor

Ziyad S. Haidar is a Full Professor of Biomaterials and Tissue Engineering and the Scientific Director of the Facultad de Odontología, Universidad de Los Andes, Chile. Prof. Dr. Haidar is the Founder/Head of the Biomaterials, Pharmaceutical Delivery and Cranio-Maxillo-Facial Bio-Engineering Laboratory/Research Group (BioMAT'X I+D+I Chile - HAiDAR R&D&I), Head of Innovation at Centro de Investigación e Innovación Biomédica (CiiB), and a Faculty Member in the Doctoral Program (BioMedicine) at the Facultad de Medicina, UANDES. He is also a Visiting Professor at several institutions, including the Division of Cranio-Maxillo-Facial Surgery, Universidad de la Frontera (UFRO)-Temuco, Chile. He is an oral and cranio-maxillo-facial surgeon with a Ph.D. in Nanobiomaterials, Pharmaceuticals, and Bioengineering from McGill University, Canada, and completed a post-doctoral training residency at the Montréal Shriners Hospital, MUHC, Montréal, QC, Canada.

Contents

Preface

This book presents innovative solutions and strategies for tackling acute and chronic problems and conditions of the oro-dental and cranio-maxillo-facial complex and beyond. Moving and/or implementing bench-top research, development, and innovation (R&D&I) to market and/or clinical practice (i.e., translation) is a lengthy and tedious process, often requiring years. As such, this book addresses the field(s) of biomimetics, bio-design, and bio-inspiration, relatively new areas of science and R&D&I where researchers look to mimic (or rather emulate) aspects of nature to help solve human problems and presents biomimicry in a simplified and practical manner to the interested reader. Back in 1903, the Wright brothers (1867–1948) were inspired by the wings of eagles to create a powered airplane and succeeded in the first-ever human flight. Today, *biomimicry* can be considered the most advanced process of applying cellular and biological principles that underlie morphology, structures, and functionality of physiological entities to design and formulate or fabricate humanmade solutions for persistent challenges. *Biomimetics - Bridging the Gap* presents biomimicry processes as either problem- or solution-based. That is, either from design to bio or from bio to design, bridging the aforementioned "gap"; a paradigm shift, perhaps. Over the past decades, R&D&I scientists (in medicine, dentistry, pharmacy, engineering, aerospace, architecture, electronics, and optics, among other fields), found renewed attraction in wonders of nature, not limited to eagle wings, lotus leaves, spider's silk and webs, fireflies, humpback whales, mother-of-pearl, lotus leaves, blue butterfly wings, Gecko's feet, and so on, as design models to help formulate or engineer optimal therapeutic solutions and devices. Intriguingly, this interest found and established ground in contemporary chemistry and materials science as well as in R&D&I fields incorporating nanobiotechnology, drug delivery, cell and gene therapy, tissue engineering, regenerative medicine, and nano-dentistry. Evidently, biomimicry has been proposed to help understand the structural and functional (physico-chemico-mechanical) properties of various biological components like cytokines, proteins, amino acids, and phospholipids to help develop numerous innovative nanotechnologies as the basis for clinical solutions. Layer-by-layer self-assembled and carbohydrate-, protein-, or peptide-functionalized lipid- or metal-based nanoparticles and nanocapsules are fine examples. Briefly, the term "*biomimetics*" originated from the Greek "*bios*" (life) and "*mimesis*" (to imitate), where, from ancient times, and for the betterment of humankind and quality of life, it is well known to adapt ideas and inspirations from nature and the surrounding environmental natural phenomena and various creatures like birds, animals, plants, and insects. Today, don't we also often tend to look for inspiration around us, or perhaps for useful information in a book, or in a famous quote or even through looking back to what some of the popular philosophers once thought and spoke? Yet, what about nature? Isn't nature with us, here, every day and a *vital* element of almost every detail of our human life? So, what lessons can be learned ... to live better? This book addresses questions such as: How does biomimicry differ from bio-design and bio-inspiration? Further, can we really, still, practically transfer nature's *lessons*, *designs*, and *inspirations* to technologies helpful or useful to humans? What is the *original* frame of reference?

These are the highlights of this book, taking the reader on a journey to the interesting world of biomimetics. This volume touches upon various technological themes to better understand how to *mimic* nature practically. It examines the design and characterization of natural and synthetic materials and the effect of biomimetic composites and coatings, 3D micro-architecture, and bio-stimulation for tissue engineering and regenerative medicine. It includes a chapter on bio-inspired tissular engineering for regenerative oral, dental, and maxilla-facial solutions, with a special focus on methodologies and processes to guide the design of new ideas and creation of novel materials, functions, devices, and therapies. Another chapter tackles the issue of biomimetics for health care, highlighting some of the most relevant, impactful, and recent innovations inspired by nature. Other chapters cover a wide spectrum of topics, such as the different aspects of the development of composites for bone tissue engineering, 3D scaffolds, and hydroxyapatite coatings with enhanced performance and bioactivity, including investigations and discussions of material surface–cell interactions and visco-elastic mechanical equivalent models. Additionally, there is a chapter on bio-simulation for inducting, via non-ionizing electromagnetic radiation, forced resonance mechanical oscillations into virus particles as a potential anti-viral modality.

This book includes four sections on various perspectives of biomimicry in medicine, dentistry, pharmacy, and materials engineering. Section 1, "Introduction to Biomimicry", provides an overview of biomimicry, from its historical basis to modern and contemporary practice. In other words, describing the journey from the drawing pad/notebook to laboratory bench-top to clinical bedside. The section begins with a chapter discussing bio-inspired tissular engineering strategies for regenerative oral, dental, and cranio-maxilla-facial clinical solutions, with subsequent chapters discussing the role and potential impact of biomimicry and bio-inspired innovations on our health, wellbeing, longevity, and quality of life. Throughout, the section presents methodologies and processes to aid in the bio-design of novel ideas and formulation of innovative materials, functions, devices, and therapies.

Section 2, "Biomimetic Tissue Engineering", dives into the design and processing details pertinent to biomaterials, where the design and characterization of natural and synthetic soft polymeric materials with biomimetic 3D microarchitecture for tissue engineering and medical applications are discussed. Then, details of novel biomaterial composites and hybrids for bone tissue engineering and osteogenesis are provided. Accordingly, it can be stated that bio-design is used in the design of artificial devices, structures, and systems in the field of bioengineering. A R&D&I sub-field that takes nature as an example and aims to make sustainable, safe, stable, functional, and durable end products. Contemplating the recent trends, the chapter includes a discussion of recent examples to raise awareness of bio-inspired and bioengineered materials.

Section 3, "Biomimetics at the Nanoscale," presents various aspects of the design, characterization, evaluation, and application of biomimetic polymers, composites, and coatings. A chapter is dedicated to assessing the incorporation or addition of fluorapatite into hydroxy-apatite coatings of prosthetics. It also discusses implementation pre-clinically. Another chapter presents the bio-simulation of inducting forced resonance mechanical oscillations (acoustic-mechanical oscillations) to virus particles by illumination using non-ionizing electromagnetic radiation. The author assumes that the viral particle is spherical in shape and then studies and analyzes the microwave signals (achieving maximal energy transfer from the microwave radiation to acoustic oscillations to the particle) theoretically and discusses the prospects of the work as a *liable* anti-virus modality. Indeed, the author suggests that this technique is feasible to compete with virus epidemics, either for the sterilization of spaces or for future therapeutic applications.

Section 4, "Biomimetic Mechanicobiology", deals with the behavior of biomaterials and investigates the strain (induced) associated with the applied stress to understand both creep and stress relaxation behavior of loaded polymeric components. Herein, via establishing the so-called mechanical equivalent models from the simple elastic element (spring with a modulus of elasticity E), simple viscous element (damper or dashpot with fluid viscosity η) to the Maxwell model, Voigt model, modified Maxwell model, modified Voigt model, and Maxwell–Voigt model, the induced strain allied with the applied force on a polymeric material is investigated and carefully discussed.

This book also proposes new examples and theoretical models for biomimetic systems and presents various applications of biomimetic materials, strategies, and solutions. *Biomimetics - Bridging the Gap* is aimed at the researcher (engineer, physicist, chemist, and biologist) interested in bio-design, bio-inspiration, bio-mineralization, biochemistry, bio-kinetics, and biomimetic solutions. It is also relevant to pharmacists, doctors, dentists, and surgeons.

> *"Biomimicry is innovation inspired by nature ... is ... the conscious emulation of life's genius."*
>
> – Janine M. Benyus; author who coined the term "biomimicry".

Dr. Ziyad S. Haidar, DDS, Cert Implantol, MSc OMFS, FRCS (C), FICD, FICS, MBA, Ph.D.
Professor,
BioMAT'X (HAiDAR R&D&I/I+D+i LAB),
Centro de Investigación e Innovación Biomédica (CiiB),
Faculties of Dentistry and Medicine (*Cross-Appointment*),
Universidad de los Andes,
Las Condes, Santiago de Chile

Section 1

Introduction to Biomimicry

Chapter 1

Introductory Chapter: bioMimetics for HealthCare – Innovations Inspired by Nature

Ziyad S. Haidar

1. Introduction

This chapter collectively aims to imitate the biological processes or systems in which nature solves problems or tackles tasks, using our world as a source of inspiration and as a guide in the design and development of new biomaterials and solutions—*innovations*.

Therefore, biomimicry is a fusion of approaches requiring the detection, perception, observance, identification, and detailed study of systems and organisms within nature, in order to use as bio-inspired models (basis) for novel scientific, technical, and technological solutions, suitable for interventional application in dentistry and medicine, including tissue bio-engineering. Such a journey also invites elucidating the underlying mechanisms and relationships between the structure and function of the stimulating natural system(s) for applied and/or translational biomimesis, biomimicry or biomimetics, to us humans, whether as sole individuals or in groups and when healthy end-users/-consumers or patients needing therapies [1].

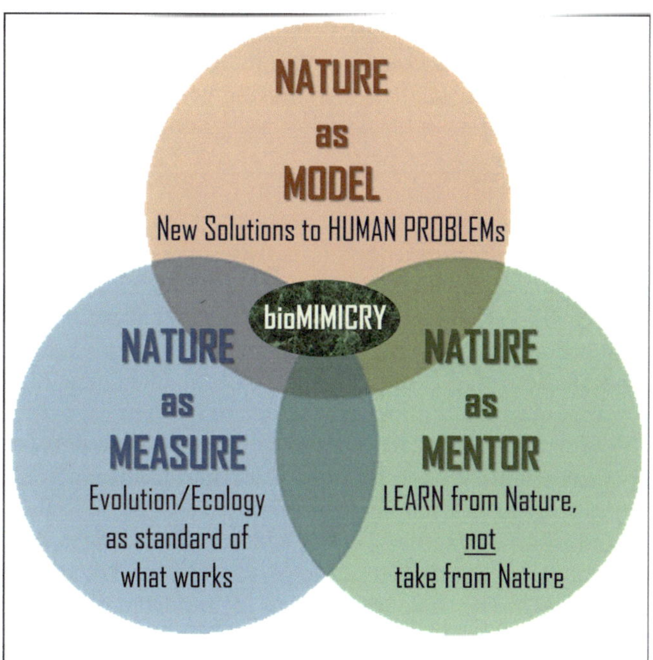

IntechOpen

Since Janine Benyus published her dominant book in 1997 [2], it can be stated that biomimicry has experienced the swift rise in the attention we have been witnessing. She outlined three essential or fundamental components for sustainable biomimicry—*active study of natural systems as source (unlimited supply) of inspiration for new bio-solutions*: (1) nature as model; (2) nature as measure; and (3) nature as mentor. Biomimicry can thus be depicted as an art form based on science, accord, and purpose.

It can also be stated, today, that in R&D&I (research, development, and innovation), the biomimicry or bio-inspired approach (mindset or point-of-view) has thus far contributed to how an investigation is conducted by pointing (directing or guiding) the way towards a more sustainable practice and future. It is noteworthy perhaps herein that biomimetics is not sufficient by itself to translate its inspiration and lessons from nature to operational devices, solutions, or technologies. Hence, does not replace disciplines and specialties such as medicine and dentistry. Rather, biomimetics and biomedical engineering, for instance, need to interplay alongside chemistry, biology, and physics, among other scientific fields, to lead to real applications that impact and benefit human-kind and our patients: *a symbiotic relationship similar to the co-existence and harmony of/between humans and nature*. If, and when realized, biomimesis and its products can manipulate the WORLD [3–5].

2. Medical bioMimicry market: market size, present trends and forecasts

According to the most recent 130-page study by Global Market Insights Inc. (August of 2021 [6]), a global market research and consultancy service headquartered in Selbyville Delaware-USA, the medical biomimetics market size is anticipated to record a valuation of USD$ 53 billion by 2027, driven by the boost in adoption of new techniques by the industry players for the provision of advanced devices to end-users. This increase or expansion is despite the losses incurred by the COVID-19 pandemic. The report provides penetrative insights presented to aid in strategic decision-making, highlighting the major trends that are likely to transform the medical biomimetics market landscape in the coming years, mainly attributed to the increasing burden of numerous healthcare challenges requiring novel solutions. Amongst those, *Tissue Engineering and Regenerative Medicine* (as a segment) is anticipated to grow at a startling CAGR of 7.5% between the period 2021 and 2027. Furthermore, the medical biomimetics industry is anticipated to continue its growth worldwide, also due to longevity and the increasing senior or geriatric population, and the surging prevalence of neurological, cardiovascular, and orthopedic diseases.

Another is the *Drug Delivery* segment, exceeding USD$ 6.5 billion in 2020, of the medical biomimetics market, mainly attributed to the increased demand and growing adoption of 3-dimensional biomimicry-based models and nano-Carriers (including nanotechnology-scaled/enabled vaccine release) in controlled drug delivery systems.

In *Dentistry*, particularly, biomimetic products that aid in (a) battling against oro-dental biofilm formation and (b) treating tooth/teeth disorders (conservative and ultra-conservative dentistry for the prevention and treatment/restoration of intact natural dentition) from decays/caries, gingival and periodontal diseases, pulp conditions and trauma/fractures, to list a few, is, also projected to considerably grow from the accounted > USD$ 3.5 million (*segment*) of the medical biomimetics market, in 2020. Interestingly, the market study [6] reported on the significance and impact of the recent advancements in the field of nanotechnology, expanding in the healthcare sector, as the demonstrated dynamic biomimetic behavior rendered feasible for exemplifying single cells at high output, concluding that such scientific, technical, and technological developments will open the door wider for nano-biomimetics (novel material composites and pharmaceutics) boosting the Medical Market, in the future.

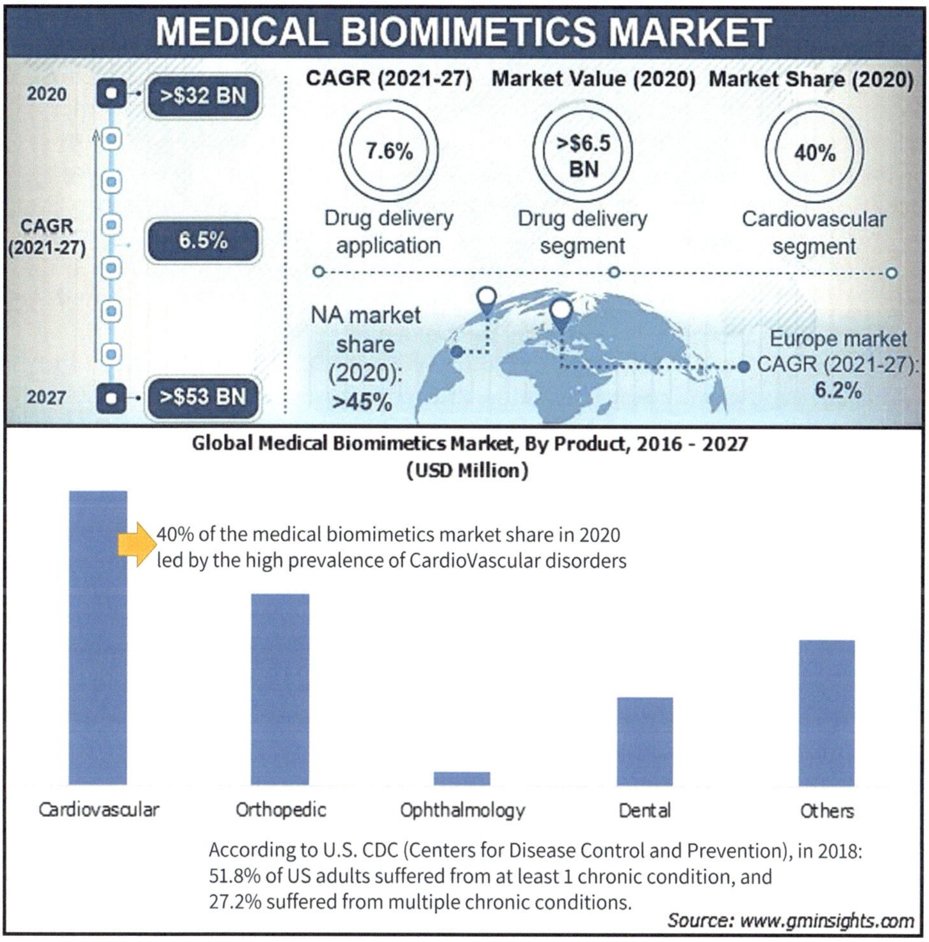

Prominent major players and competitors operating in the medical biomimetics market include Abbott, BioHorizons, Biomimetics Laboratories Inc., BioTomo Pty. Ltd., Avinent Implant System, Hstar Technologies corporation, and Veryan medical among others. Herein, collaborations, partnerships, and acquisitions are among the various business strategies recently visibly adopted to enhance the co-market share [6]. Finally, the clinical and regulatory requirements for approving the translation of biomimetic products continue to present a stringent hurdle and challenge to them.

3. Nature as R&D&I laboratory: examples of biomimetic HealthCare

Nature retains an unlimited imagination. Japan's famous high-speed 240–320 km/h *Shinkansen bullet* train [7] was inspired by the shape of the Kingfisher's beak (*Alcedinidae*), a small to medium-sized brightly colored bird with long, narrow-pointed, and dagger-like beak allowing to dive into the water without splashing to catch prey. Not only did this bio-inspiration help to reduce noise and eliminate tunnel booms, yet also allowed the train to travel 10% faster using 15% less electricity. A US-Canadian team (finalists for the European Inventor Award in 2018) innovated turbine blades with three-dimensional bumps [8] on their leading edges based on viewing the "bumpy" flippers (tubercles) of humpback whales (*Megaptera novaeangliae*). They discovered that this helped the 14–18 m long and ~40 metric ton fish reduce unwanted whirling masses of air (vortices), thereby reducing drag while simultaneously increasing lift. The team's improved turbine aerodynamic performance can help wind farms generate up to 20% more power and increase airflow by up to 25% in industrial fans and blowers, whilst producing less noise by at least 2 decibels and requiring less maintenance (life-time of wind turbine increased by 25%). Henceforth, and given such competitive benefits (~20% better) to market leaders, their first licensed product is forecasted to be worth ~USD 10 billion in 2022.

In HealthCare, studies in biomedical technology, bioengineering, and dental biomaterials, for example, have already shown that our science and engineering cannot currently *out-perform* many of nature's capabilities. The COVID-19 pandemic reminded us that we are constantly vulnerable to life-threatening invasions from bacterial species, many of which have existed on the planet billions of years before us. Biomimicry [9, 10], the study of the formation, structure, and function of biologically produced substances and materials, mechanisms, and processes, as stated earlier, can help design, develop, formulate, fabricate and translate new biosolutions.

For example, Biomatrica [11], a subsidiary of Exact Sciences Corporation (following acquisition in 2018 for USD$ 20 million), a Wisconsin-based molecular diagnostics and cancer screening company, to overcome the problem that many vaccines are lost due to breaks in refrigeration during shipping and treatment (a hurdle we all witnessed during the Coronavirus infectious disease or COVID-19 caused by the SARS-CoV-2 virus pandemic), adapted and incorporated into their product a process (ambient temperature storage reagents for DNA and RNA) inspired from the Tardigrade, a millimeter-long cousin of the arthropods. Briefly, these creatures employ a protective process called *anhydrobiosis*, which safeguards their DNA, RNA, and proteins until water revives them, even though they can dry out for up to 120 years. Basically, the tardigrade releases trehalose, a simple sugar molecule and as the water leaves their cells, the trehalose replaces where the water once was, and the cell membrane releases the water and bonds to the sugar instead, hence, by doing this, the proteins stay in the same place that they would be when fully hydrated. The company used this bioinspiration to protect live vaccines so that they no longer need to be refrigerated. Such can be helpful for vaccinating vulnerable populations in tropical areas. SB 3000 [12] or Swedish Biomimetics 3000, a ground-breaking life science and pharmaceutical company headquartered in Copenhagen-Denmark, found inspiration in the defense mechanism of ground carnivorous *Bombardier* beetles (*Carabidae*) to develop a micro-mist spray technology with potential application in nebulizers (a type of breathing machine that lets you inhale medicated vapors), called μMIST, that has a lower carbon print/impact than aerosol sprays, as it does not require a propellant to work (spray highly-viscous formulations). Those beetles, when disturbed, repel attacking insects by ejecting a hot (near-boiling temperature) noxious chemical spray produced (via a chemical reaction between hydroquinone and hydrogen peroxide) in their abdomen (alongside a popping sound). At Kansai University in Osaka-Japan [13], a team of engineers found inspiration in mosquito bites, replicating their *proboscis*, to develop pain-free needles and injections, to replace the conventional hypodermic steel needles, that despite being smooth, do penetrate deep and leave ample metal contact with skin tissue hence causing us pain. Studies revealed that mosquitos inject us by vibrating their proboscis to help the serrated sections of their maxillae ease down, with least sensation possible, through our skin. We do not feel the bite itself because this small-/close-contact mechanism reduces friction and in consequence, nerve stimulation, but feel discomfort *afterward* because the mosquitos inject bacteria that cause irritation and pain [13].

For bioprinting, a novel bio-ink incorporating Hyaluronic Acid (HA), a natural linear polysaccharide found in many tissues throughout the human body, was recently developed by Rutgers University in New Jersey and one of America's leading public research universities [14]. Herein, the HA bio-polymer, whose main function is to retain water to keep tissues well lubricated and moist, is also well known to play an important role in regulating cell differentiation, migration, angiogenesis, and inflammation/immunological responses. Briefly, the bio-ink material is made of modified HA and polyethylene glycol to serve as the basic "ink cartridge" for the 3-D printing of different scaffolds (in a range of physico-chemico-mechanical/rheological and biological properties; personalized/customized design and manufacturing) that can be employed for growing, restoring and replacing the lost and/or defective human tissues, overcoming few of the main challenges in the field of 3-D bio-printing.

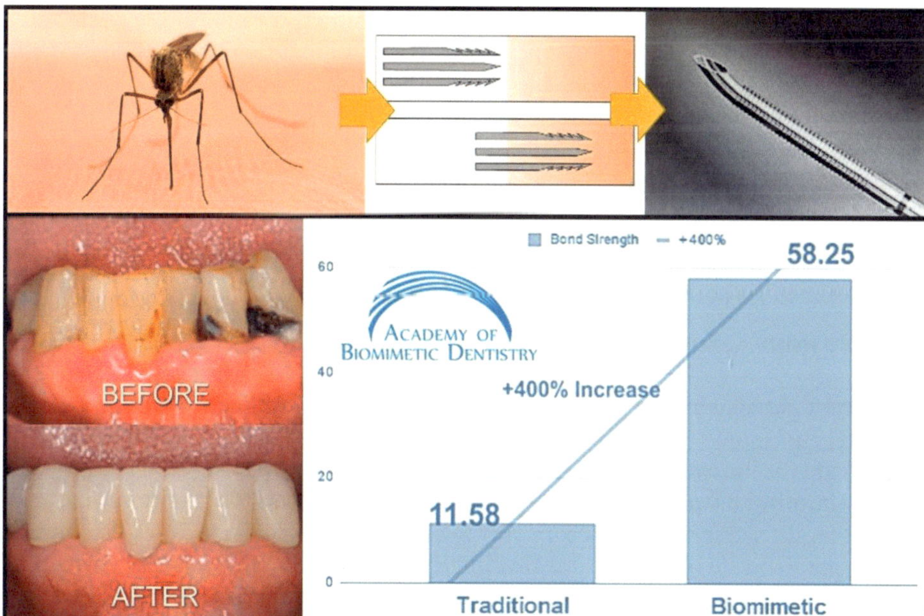

ABOVE: Pain-less Needle for Injections inspired by Maxillae of Mosquitos.
BELOW: bioMIMETIC DENTISTRY and the adhesive (restorative material) dental bond achieved with 400% increase in strength (equivalent to strength of bond between natural enamel and dentin).

Inspired by squid [15], a *mollusc cephalopod* with an elongated soft body, large eyes, eight arms, tentacles, and more particularly the teeth present in ring formations inside suction cups on those tentacles, researchers Penn State University in Pennsylvania-USA, developed a self-assembled composite material with tunable electric properties for bio-engineering use. Herein, they noticed that the squid ring teeth are made up of proteins that can combine (or assemble) in different ways to help the squid grip onto a surface or grasp prey. Also, if the teeth break, they can self-heal. The structure of the squid ring teeth proteins (tandem repeat proteins) helped inspire a solution to problems in creating mixtures (matrix-to-filler ratios, in tiny or small areas) with highly-tunable properties, suitable for improving electronic devices, such as diodes or regulators, as well as topologically-networked biomaterials to isotropically and anisotropically modulate the "electronic transport" in composites.

Biomimetic Dentistry (including regenerative dentistry) [16], on the other hand, is the art and science of clinically restoring damaged teeth and using materials that mimic the properties of natural teeth—in terms of strength, appearance, and function (main reference to adhesive materials in restorative dentistry). Indeed, according to the Academy of Biomimetic Dentistry [17], the biomimetic dentistry approach (and currently-available materials) can help dentists conserve as much of the tooth structure as possible, preserve tooth vitality and prevent unnecessary damage to root canals, restore teeth that would otherwise (traditionally) need an extraction, increase the bond strength of dental restorations by 400%, minimize shrinkage stress on the teeth, eliminate sensitivity/pain, and create long-lasting restorations that prevent complications often experienced with conventional approaches. 3-D printing might also be helpful for the next generation of bone grafts to clinically-create on-demand patient-specific scaffolds [18, 19]. Indeed, in the developing area of regenerative dentistry, the "bio-tooth" [20], perhaps is a fine biomimesis or biomimicry example, *in pursuit*. Herein, for a bio-engineered and fully-functional tooth, the *synergistic* employment of the accruing understanding of the underlying cellular and molecular biology into the creative (and patient-specific) design of oral, dental (and cranio-maxillo-facial therapies) that ultimately aim to restore, repair, rejuvenate, replace and/or regenerate defected, damaged/injured or lost tissues is *key*. Collectively, alongside incorporating tissue engineering and regenerative nano-biomaterials would boost and expedite high-quality research to eventually realize it.

4. Conclusions

The examples illustrating biomimicry for health and in novel healthcare applications are plenty, and some will be explored within the chapters of this book, to further demonstrate the innovative form of biomimetic technology that imitates (or mimics) nature to improve human lives via creating desirable solutions. To re-emphasize, such a process requires the study of nature and natural phenomena, principles, and underlying mechanisms, to obtain bio-inspiration that may benefit various applied scientific and technological disciplines. Smart/Intelligent nano-biomaterials for Tissue Engineering, Regenerative Medicine and Regenerative Dentistry is a fine example. It is also perhaps worth mentioning in this *introductory chapter* that biomimicry can go above and beyond the *simplistic* bio-inspiration and use of natural properties as the basis for innovation and translation of new products to the demanding market of end-users and patients. It can bridge the gap between the lab and the industry via the intra-disciplinary design and formulation of functional solutions combining knowledge, methods, techniques, and advances in the fields of chemistry, biology, architecture, engineering, medicine, pharmaceutics, and dentistry, alongside contributions from artificial intelligence, robotics, bio-informatics, and omics. Indeed, biomimesis, today, can be considered the leading scientifically-relevant paradigm for innovative design and the guide for advancing new methods and devices, for a higher scientific, technological, technical, medical, and socio-economic impact. There is still much to learn from nature and the world around us, including ourselves since we are one of the most successful end-products and -users.

Acknowledgements

This work was supported by operating grants provided to BioMAT'X I+D+I (HAIDAR LAB: Laboratorio de Biomateriales, Farmacéuticos y Bioingeniería de Tejidos Cráneo Máxilo-Facial), a member of CiiB, Faculty of Medicine, Universidad de los Andes, through the awarded project funds: (1) NAM-USA/ANID-Chile # NAM 21I0022 (2021-2023) and (2) CORFO Crea y Valida # 21CVC2-183649 (2022-2024).

Conflict of interest

The author declares no conflict of interest.

Notes/thanks/other declarations

I would like to thank Mr. Josip Knapić, Author Service Manager at IntechOpen for his constant support, communication, feedback and in the timely-preparation of our project "Biomimetics—Bridging the Gap", with content, to the best of abilities, *different* from what has been previously published. Thank you and hope for the next.

Author details

Ziyad S. Haidar
Faculties of Medicine and Dentistry, BioMAT'X (HAIDAR LAB) R&D&I, Biomedical Research and Innovation Center (CiiB), University of the Andes, Santiago, Chile

*Address all correspondence to: zhaidar@uandes.cl; zhaidar78@gmail.com

References

[1] Benyus JM. Biomimicry. HarperCollins e-Books; New York, USA. 2009

[2] Benyus JM. Biomimicry: Innovation Inspired by Nature. New York: William Morrow and Co; 1997. p. 308

[3] Ilieva L, Ursano I, Traista L, Hoffmann B, Dahy H. Biomimicry as a sustainable design methodology—Introducing the 'Biomimicry for Sustainability' framework. Biomimetics. 2022;**7**(2):37

[4] Perera AS, Coppens MO. Re-designing materials for biomedical applications: From biomimicry to nature-inspired chemical engineering. Philosophical Transactions. Series A, Mathematical, Physical, and Engineering Sciences. 2019;**377**(2138):20180268. DOI: 10.1098/rsta.2018.0268

[5] Haidar ZS. Bio-inspired/-functional colloidal core-shell polymeric-based nanosystems: Technology promise in tissue engineering, bioimaging and nanomedicine. Polymers. 2010;**2**:323-352

[6] Report ID: GMI767 Published Date: Aug 2021-Medical Biomimetics Market Size By Product (Cardiovascular, Orthopedic {Prostheses, Exoskeleton}, Ophthalmology, Dental), By Application (Plastic Surgery, Wound Healing, Tissue Engineering, Drug Delivery), COVID-19 Impact Analysis, Regional Outlook, Application Potential, Competitive Market Share & Forecast, 2021-2027. Available from: https://www.gminsights.com/toc/detail/medical-biomimetics-market

[7] Shinkansen-The Japanese bullet trains. Available from: www.jrailpass.com/shinkansen-bullet-trains and asknature.org/innovation/high-speed-train-inspired-by-the-kingfisher/

[8] Patent # US20090074578A1Turbine and compressor employing tubercle leading edge rotor design. Inventor: Stephen W. Dewar, Phillip Watts and Frank Eliot Fish. Available from: https://worldwide.espacenet.com/publicationDetails/biblio?DB=EPODOC&II=0&ND=3&adjacent=true&locale=en_EP&FT=D&date=20070711&CC=EP&NR=1805412A1&KC=A1

[9] Upadhyay A, Pillai S, Khayambashi P, et al. Biomimetic aspects of oral and dentofacial regeneration. Biomimetics (Basel). 2020;**5**(4):51

[10] Hwang J, Jeong Y, Park JM, Lee KH, Hong JW, Choi J. Biomimetics: Forecasting the future of science, engineering, and medicine. International Journal of Nanomedicine. 2015;**10**:5701-5713

[11] Biomatrica. Exact Sciences Corporation, Wisconsin, USA: Available from: https://www.sbir.gov/sbc/biomatrica-inc

[12] Swedish Biomimetics 3000, Copenhagen-Denmark. Available from: https://sb3000.tech/

[13] Tolfree D, Smith A. A painless bite. Physics Review. 2018;**27**(4):18-20

[14] Godesky MD, Shreiber DI. Hyaluronic acid-based hydrogels with independently tunable mechanical and bioactive signaling features. Biointerphases. 2020;**14**(6):061005. DOI: 10.1063/1.5126493

[15] Vural M, Zhu H, Pena-Francesch A, Jung H, Allen BD, Demirel MC.

Self-assembly of topologically networked protein-Ti3C2Tx MXene composites. ACS Nano. 2020;**14**(6):6956-6967. DOI: 10.1021/acsnano.0c01431

[16] Sharpe P. Regenerative dentistry. Frontiers in Dental Medicine. 2020;**1**:3. DOI: 10.3389/fdmed.2020.00003

[17] Academy of Biomimetic Dentistry. Available from: https://www.aobmd.org/

[18] Ashammakhi N, Kaarela O. Three-dimensional bioprinting can help bone. The Journal of Craniofacial Surgery. 2018;**29**:9

[19] Zhang L, Yang G, Johnson BN, Jia X. Three-dimensional (3D) printed scaffold and material selection for bone repair. Acta Biomaterialia. 2019;**84**:16-33. DOI: 10.1016/j.actbio.2018.11.039

[20] Mitsiadis TA, Harada H. Regenerated teeth: The future of tooth replacement. Regenerative Medicine. 2015;**10**:5-8. DOI: 10.2217/rmc.14.78

Chapter 2

Perspective Chapter: Biomimetics – Bio-Inspired Tissular Engineering for Regenerative Oral, Dental and Cranio-Maxillo-Facial Solutions

Ziyad S. Haidar

Abstract

This chapter introduces the scope of the book—bioMIMETICS can be described as an innovative form of technology that imitates (or mimics) nature in order to improve human lives via creating desirable solutions. It is the study of nature and natural phenomena, principles, and underlying mechanisms, to obtain bio-inspired that may benefit various applied scientific and technological disciplines. Smart/Intelligent nano-bioMaterials for Tissue Engineering and Regenerative Medicine are a fine example. Yet, biomimicry can go above and beyond the simplistic inspiration and use of natural properties as the basis for innovation of new products. It bridges the gap between the lab and the industry, via the intra-disciplinary design and formulation of functional solutions combining knowledge, methods, techniques, and advances in the fields of chemistry, biology, architecture, engineering, medicine, pharmaceutics, dentistry, and biomedical engineering. Three-Dimensional Printing, Hybrid nanoCoatings, and Stimuli-sensitive and -responsive Cell/Drug Delivery Systems, and Robotics are some of the topics covered in this new book. In this first chapter, a general overview of bio-inspired materials, technologies, and strategies, collectively known as "bioMiMETICS," is presented to bridge the gap between the laboratory "bench-top" and translational application, particularly, the clinic or "bed-/chair-side," with a focus on "REGENERATIVE DENTISTRY" and the "CRANIO-MAXILLO-FACIAL bio-COMPLEX."

Keywords: biomimetics, dentistry, tissue engineering, regenerative medicine, surgery, bioprinting, scaffolds, hydrogels, nanotechnology, biomaterial innovation

1. Introduction

The Greek words "bios" and "mimesis", for (life) and (to imitate), respectively, from the term "biomimetics," hence, *biomimicry*, coined by Otto Schmitt, in 1957, can be now described as an innovative form of creative thinking, design, and technology that uses or imitates (or mimics) nature to improve human lives via creating desirable solutions and devices [1–3]. The paradigm idea of seeking inspiration from nature, the center ground or concept for biomimetics, is and cannot be thought of as recent field of study, discovery, and impact. It can be stated, that in R&D&I (research, development, and innovation), the biomimicry or bio-inspired approach (mind-set or point-of-view) has thus far contributed to how investigation is conducted by pointing (directing or guiding) the way toward a more sustainable practice and future. It is noteworthy perhaps herein that biomimetics is not sufficient by itself to translate its inspiration and lessons from nature to operational devices, solutions, or technologies. Hence, does not replace disciplines and specialties such as medicine and dentistry. Rather, biomimetics and biomedical engineering, for instance, need to interplay, alongside chemistry, biology, and physics, among other scientific fields, to lead to real applications that impact and benefit humankind and our patients (**Figure 1**). A symbiotic relationship similar to the coexistence and harmony of/between humans and nature. If, and when realized, biomimetics and its novel products can manipulate the WORLD [1–7]. bioMIMICRY is an art form based on science, accord, and purpose.

Indeed, biomimetics *perhaps* corresponds to the consideration and comprehension (and inspiration by) of surrounding natural structures and functions, herein, of cellular and biological systems, alongside the conforming translation of the observed operative or operational principles as fundamental models for the creative design and development of novel technical systems with *further* enhanced properties [8]. Natural

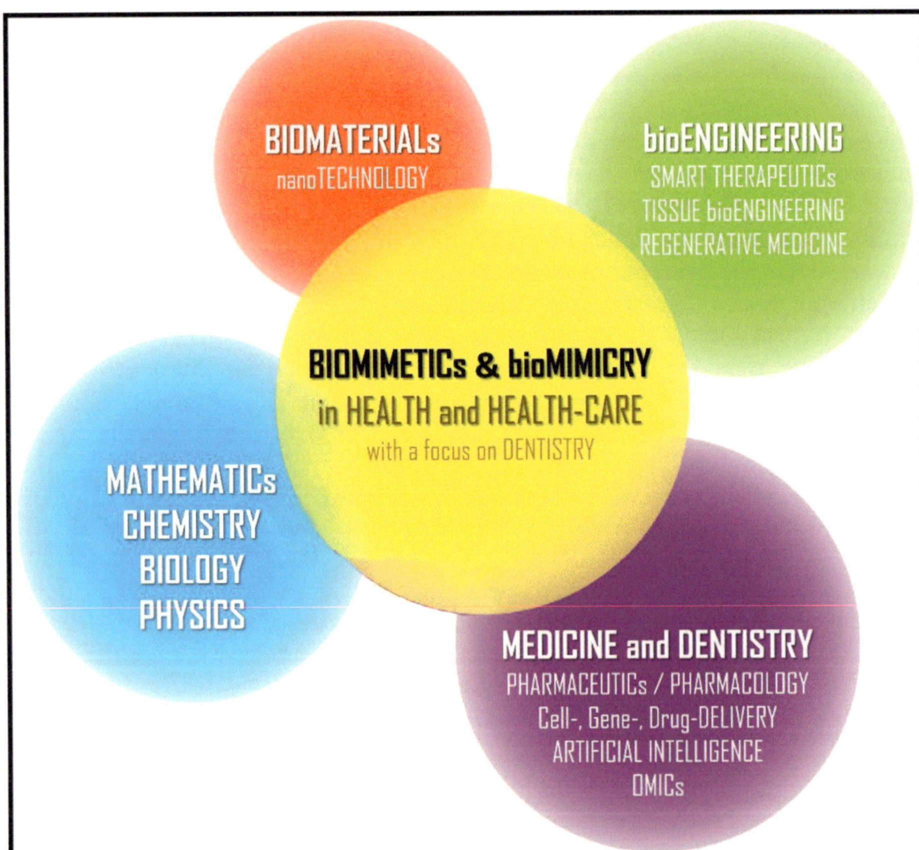

Figure 1.
Inter-/Intra-/Multi-Disciplinary interplay for bioMIMETIC Health Care.

structural features have played a role in the evolution and enhancement of specific intrinsic material properties, later, providing ground for numerous technical applications and tools in architectural design and construction, advanced biomaterials, medicine and robotics, surface engineering and bio-coatings of medical and dental implants, to list a few examples [8–10]. Such, when supplemented with the accruing study of cellular (+ mesenchymal stem/stromal cell) behavior, interactions and communications alongside cell signaling and creating a controlled or adequate cell environment, tissue engineering, and regenerative medicine advances [1, 2, 9–12].

In this first introductory chapter to the book, a general overview of bio-inspired and biomimetic materials, technologies, methods, and strategies is presented to bridge the gap between the laboratory "bench-top" and translational application, particularly, the clinic or "bed-/chair-side," with a focus on DENTISTRY, in general, and the sub-specialty areas of Oro-Dental, Cranial, and Maxillo-Facial bio-Engineering.

Hwang *et al.* in 2015 [8] mentioned that biomimetics has a long history extending from knives and axes inspired by the dental structures of currently extinct animals to today's strongest cutting-edge Carbon nanoMaterials employed in bioEngineering, passing through Leonardo da Vinci's "flying machine" inspired by a bird and the Wright brothers' powered airplane and the first *successful* human flight, back in 1903 [4, 5]. Hence, it has been noted, suggested, and/or suspected that *biomimicry* might

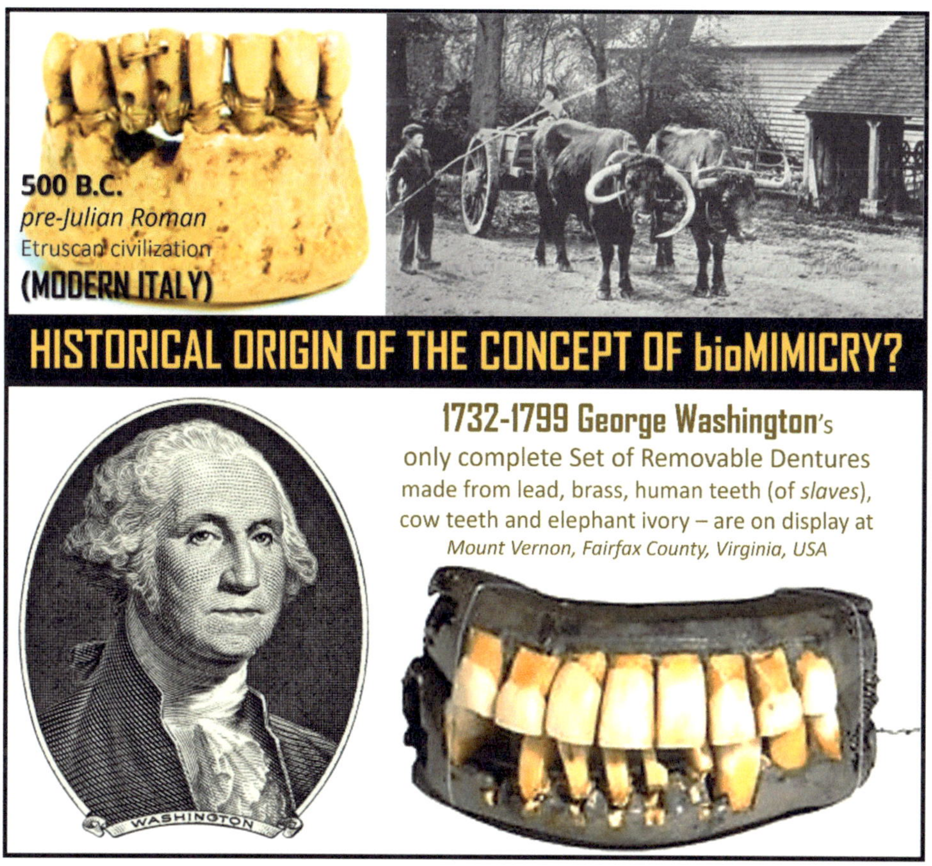

Figure 2.
bioMIMETICs date back to > 2500 years ago with carved OX-bone teeth.

actually date back to more than 2500 years ago, when artificial teeth were carved from the bones of oxen (ox or bullock), as the first attempt to replace body parts [8]. **INSPIRING!** Indeed, the substitution of natural materials by artificial biomaterials that *mimic* natural (original) tissue and tissular structure and function is yet another biomimetic approach; in fact, it is more prominent in tissue engineering and regenerative medicine. The idea of dental Dentin/Pulp renewal is a fine example (**Figure 2**) [13].

2. Fibonacci's sequence, ratio, and THE GOLDEN SPIRAL in bioMIMICRY

In 1997, the book "Biomimicry" by Janine M Benyus was published [14], suggesting (and even emphasizing) that biomimicry, via absorbing lessons from nature (as the groundwork for products rather than just a source for raw materials), is leading the path toward a new era of technological development. Indeed, biomimetic technologies arise from an inter-/intra-disciplinary flow of ideas, benefiting from the millions of years of creative design effort achieved by natural selection in living systems. Benyus, in 2002 (eBook in 2009) published a new book on how innovations inspired by nature are rapidly transforming the life on earth (**Figure 3**) [15].

Hence, for a practical, convenient, operational, and/or translational (and profitable) bio-inspired design or innovation (reaching the end-user: consumer or patient for example), the biologically inspiring natural system or organism is to be studied and understood, rather than simply copying, fusing, or mimicking the creative design

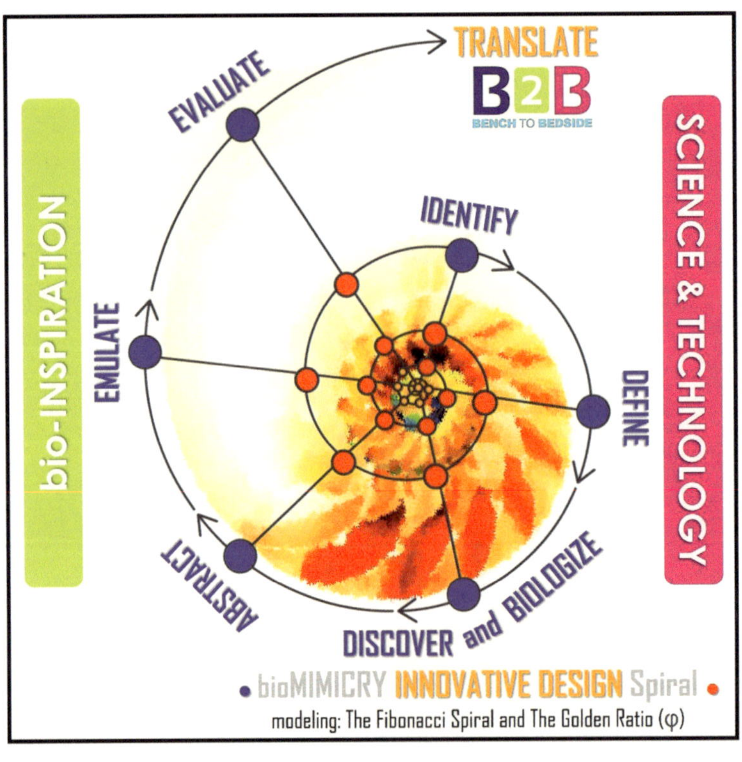

Figure 3.
bioMIMICRY SPIRAL: IDEA to MARKET, following the "Golden Ratio."

by itself [8, 14, 15]. In a study on how rabbits reproduce, a thirteenth-century Italian mathematician Leonardo Bonacci (later known as Leonardo Fibonacci) was the first to write (*Liber Abaci*, published in 1202) about the sequence found in nature and in the world around us, later referred to as the Fibonacci Sequence [16–18]. Basically, the sequence begins with 0 and 1, then continues with the sum of the two preceding numbers: 0, 1, 1, 2, 3, 5, 8, 13, 21, 34, 55, 89, 144, 233, 377, 610, 987, 1597, 2584, 4181, and so on (the sequence follows the *golden* rule that each number is equal to the sum of the preceding two numbers; *i.e.,* each number is approximately 1.618 times greater than the preceding number), leading to calculating/applying (hidden inside the sequence) the *golden* ratio (phi = φ) and employing the *golden* spiral (a logarithmic spiral with a shape that is infinitely repeated when magnified, via A using quarter-circle arcs inscribed in squares generated from the Fibonacci sequence) in ordering and quantifications (and DATA-bases) [14]. To simplify further, φ is defined as the ratio of a rectangle with dimensions A x B where the ratio A/B is equal to (A + B)/A; regarded by many artists as the perfect proportion for a canvas. Fibonacci explained that these numbers are at the heart of how things grow in the natural world (**Figure 4**).

Nature uses what it has grown so far to make the next move. It also helps describe *predictable* patterns on everything, from atoms and sub-atomics to huge stars in the sky [8, 14–16]. Interestingly, the ratio of the total number of chapters in the Quran (114), which represents the physical design of the Quran divided by the Quran Constant (70.44911244), which represents the mathematical design of the Qur'an

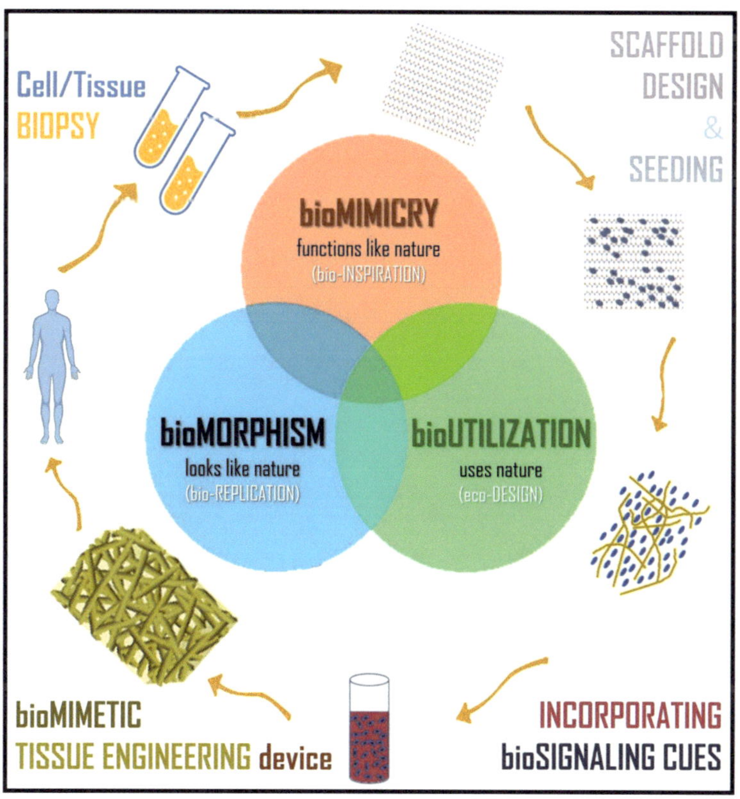

Figure 4.
Distinctives of bioMIMETICs and utility in Tissue/Organ bioEngineering.

(Koran) gives 1.6181893; it is amazingly equal to the golden ratio φ [19]. Nature uses this ratio to maintain balance, and the financial markets, for example, seem to do so as well. Fibonacci earlier explained how these numbers keep track of the population growth of rabbits. If a pair of rabbits take a month to mature before it can give birth to a new pair of rabbits, how many pairs of rabbits will there be each month? The answer is in the Fibonacci sequence [8, 14–16, 20]. Biomimicry is all about deep patterns and emulating the genius of genius to create conditions conducive to our Life; a more sustainable future. Velcro®, concentrated solar arrays and bioWAVE tidal energy are fine examples of implementing the Fibonacci Sequence, Ratio, and Spiral used to mimic the structural form and design functions of selected natural organisms for Innovative Design and re-Design [20]. Remember that nature-inspired design goes beyond structural, functional, or aesthetic similarities, and delves into the physico-chemico-mechanical and biological features of natural systems. Herein, for bio- or nature-inspired chemical, biological, medical, dental, and/or bioengineering, differentiating between (i) nature-inspired (or bio-inspired), (ii) nature-mimicking

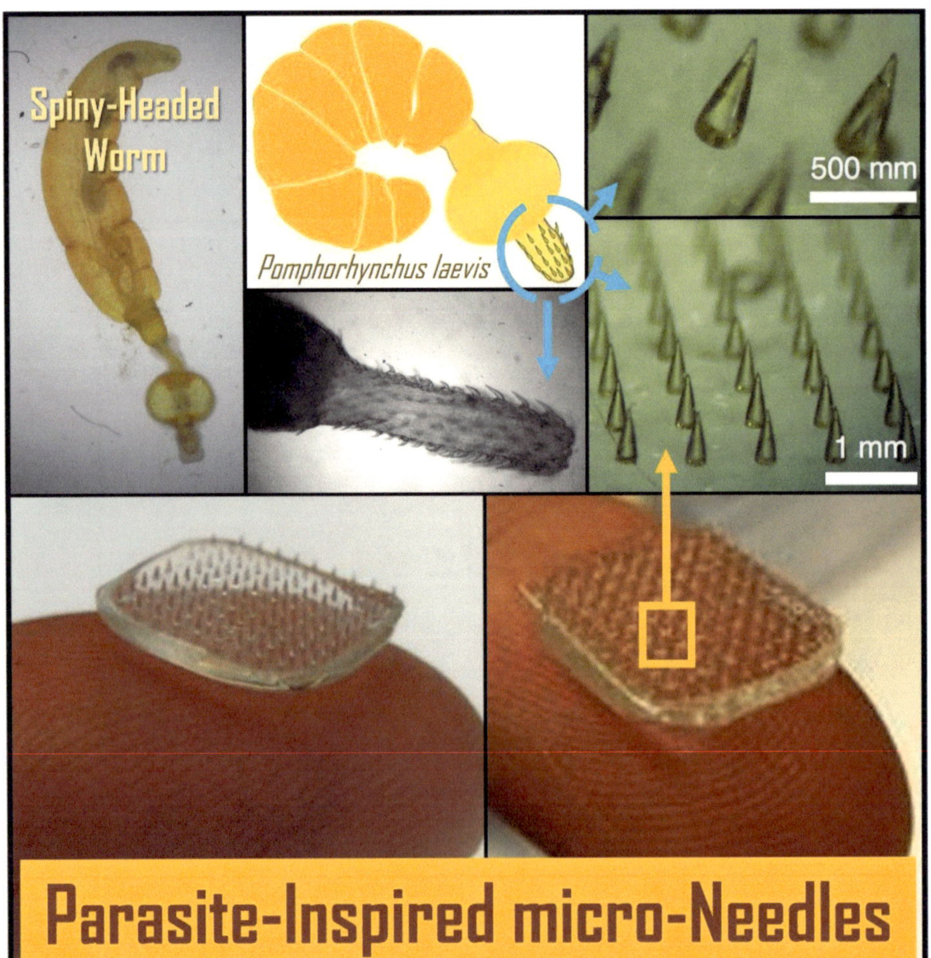

Figure 5.
Jeffrey Karp's bio-medical adhesive patch developed at Harvard Medical School and Brigham and Women's Hospital, where he is a bio-Engineer and Professor.

(or bio-morphism), or (iii) nature-integrated (or bio-integration) design, is *perhaps* critical. Perera and Coppens [21] recently examined those distinctive approaches, through illustrative examples, employed to deduce or outline a systematic methodology useful for translating innovative solutions, from the laboratory (bench-top) to market (bed-/-chair side/end-user); concept of "*nature-inspired chemical engineering*" [21]. **Figure 5** demonstrates a micro-needle/medical adhesive patch device inspired by an electron micrograph of a spiny-headed worm (*Pomphorhynchus laevis*) that lives in fish by swelling the tip of its proboscis to anchor/latches itself to the flesh once inside the gut [22].

The creative adhesive device consists of a sheet of micro-needles whose tips swell upon contact with water, with potential applications in localized and targeted drug delivery and skin grafting (including burn wounds) among other possibilities. Combining polystyrene and polyacrylic acid, the adhesive micro-needles would replace traditional staples or sutures during plastic and reconstructive surgeries via providing multiple points of contact and adherence (universal soft tissue adhesion with minimal damage), thereby shortening the operative/surgical time [22]. Further, the patch can deliver antimicrobial or anti-inflammatory drugs directly to the skin graft sites while holding them together, *in situ*, thereby reducing the risk of infection and accelerating healing time for patients.

3. bioMIMETIC/bio-INSPIRED devices, fibers, robotics, and art designs

For this introductory chapter, other illustrative examples were compiled in an attempt to provide the reader with a different view on recent biomimetic innovations.

At Delft University of Technology (TU Delft) located in the Netherlands, researchers searched bio inspiration to apply in designing novel surgical instruments and to improve how surgical techniques and procedures are performed [23]. A main driver was on finding a way to make rigid tools more naturally flexible steerable, yet without damaging nearby tissues (minimally-invasive), for use in knee arthroscopies, for example. Herein, creative mechanical engineering, inspired by snakes (**Figure 6A**) and their elasticity, helped develop such an arthroscopy tool that has a snake-like tip capable of bending, tentacle-like, yet in direct response to the hand movements of the operator/surgeon; a potential invention for robotic applications [23]. **Figure 6B** displays a *stealth* (infrared or IR) porous fibrous fabric (textile woven composed of aligned biomimetic fibers) inspired by the Polar bears and mimics the structure of their hairs [24]. Briefly, The Zhejiang University (ZJU China) researchers found that Polar hairs have a sponge-like hollow network core, which reflects back IR emissions from the animal body and helps prevent heat loss and thereby keeps the bears warm in their Arctic environments. Therefore, using fibroin (a protein in silk) and chitosan (from chitin) solution alongside a novel freeze-spinning method (developed for scale-up or large-scale production), they were able to continuously fabricate aligned, porous (87% porosity) micro-structured fibers (~200 μm wide) that are strong/sturdy, yet wearable and breathable, and highly thermally insulating, with capacity for active electro-heating if/when doped with Carbon nano-tubes, and application extending for commercial/personal insulation to military light-weight thermal management [24].

In 2022, a novel modular biomimetic live working robot was designed and developed [25], inspired by the motion of inch-worms (which have two rows of legs on head and tail to latch, stick, or suck onto objects for bodily support), for use in the

Figure 6.
Illustrating bioMIMICRY in different fields of innovation and socioeconomic impact. A. Surgical Tool inspired by Snakes and their flexible bio-mechanical movement, by TU Delft. B. Stealth and Heat-insulating Textiles inspired by the hairs of Polar Bears. C. Robotic Climbing System for Power Distribution Lines inspired by the Inch-Worm and how it moves and latches to surfaces. D. Architectural construction inspired by the shape and lattice structure of the Euplectella aspergillum.

power distribution line, is depicted in **Figure 6C**. Basically, the 3-D rendered robotic system is intended to climb up the pole and complete the live working on power distribution lines, instead of the human operator, as a safer and more effective way to improve the reliability of power supply, especially in difficult or challenging areas such as mountains, as an alternative to traditional aerial lifts. The single-body robotic system (weighing less than 25 kg) demonstrated flexibility and rapidity in pole climbing via its modular design, thereby facilitating different configuration combinations in order to achieve different movement and performance requirements [25].

Finally, although unrelated to medicine or dentistry, **Figure 6D** displays a beautiful example, one of countless around the World, of biomimicry in Architecture and construction; Norman Foster's iconic Skyscraper (of sustainable, high-tech and post-modern/neo-futurism architecture), also referred to as The Gherkin, in London's primary financial district. The building, completed in December of 2003 and opened in April of 2004, mimics the shape and lattice structure of the Venus Flower Basket Sponge (*Euplectella aspergillum*); a marine glass sponge in the phylum Porifera found in the deep waters of the Pacific Ocean (at depths ≤ 500 m). If interested in architectural design, check also the Institute du Monde Arabe in Paris, France, which mimics in its structure and façade the iris of the human eye. Irresistible to include architectural art and creative biomimetic design herein, to further emphasize that the idea of mimicking nature in man-made inventions is not new, and biomimetics is not solely (or limited to) the chemical, biological and/or medical. Furthermore, it is perhaps worth mentioning herein that the previously discussed golden ratio or Φ is also considered as one of the oldest rules in architecture, for example, in how transfer the weight of a massive building to its foundation. Indeed, the Fibonacci sequence has been used for centuries by man-kind for centuries, where architects leveraged Φ to create balance between the structural elements, as in the Pyramids of Giza in Egypt to the Parthenon Temple of goddess *Athena* on the Athenian Acropolis in Greece to the enormous Baalbeck Temple of the Phoenician sky god, *Baal*, in the Beqaa Valley of Lebanon. In essence, *biomimicry* can also be described as the learning and adapting processes from the wisdom of nature and its designs to then apply creatively and practically the inspired and acquired knowledge and sense to provide innovative and necessary solutions to help the humans and improve our World. In fact, biomimicry and biomimetics in health care can also be referred to or looked at as *architectural medicine* [26], emulating the time-tested patterns and strategies of nature to create new products, processes (and policies)—new ways of sustainable living—to challenging problems well-adapted to life on Earth, long-term.

4. bioMIMETICs advancing nano-bioMaterials, nanoTECHNOLOGY, and TISSUE ENGNEERING for bio-MEDICINE/-DENTISTRY and HEALTH

TISSUE ENGINEERING [27, 28], an inter-/multi-disciplinary approach, essentially, seeks to create tissues with optimal performance for clinical applications. Various factors, including cells, biomaterials, cell or tissue culture conditions (cell-cell and cell-material interactions), and signaling cues or molecules such as cytokines and growth factors, play a vital role in the engineering of musculo-skeletal tissues. Herein, the *in vivo* micro-environment of cells imposes complex and specific stimuli on the cells and has a direct effect on cellular behavior, including proliferation, differentiation, and extracellular matrix (ECM) assembly, as indicated earlier. Therefore, to create appropriate musculoskeletal tissues, the conditions of the natural environment around the cells should be well replicated, imitated, or mimicked [27–35]. Therefore, for engineering and creation closer to natural tissues in terms of appearance and function, researchers continue to attempt at developing biomimetic multifunctional scaffolds that can function better mechanically while producing the appropriate cellular responses in terms of cell signaling and cell adhesion. Indeed, in biomedical design, engineering principles are applied to medicine, dentistry, pharmacy, and cellular/biological systems for the objective of designing and translating novel pharmaceutical, tissue engineering, and regenerative medicine applications, methods, techniques, formulations, and tools for health care [28, 30–33]. This includes therapeutics as well as analysis and diagnostics, thereby innovating solutions to challenging health problems. Fundamentally, bioMIMICRY, through various methodologies as well as new ideas on the creation of novel materials and functions, is how to mimic nature, as described previously. Nature has given us plenty of ideas on how to build composites and organized structure. In material and biomaterial science, the development of interfaces that integrate the functions of living cells and materials is *key*. Hence, the structure of a biomaterial influences cellular response(s) thereby determining the potential biomedical application(s). Modern nano-biomaterials, for example, are combinations of the unique properties offered by the organic and inorganic constituents of/from/within a single material, on a nano-scale; nanocomposites [8, 10, 21, 28, 29].

NANOTECHNOLOGY [21, 28, 31–37], emerged from the physical, chemical, biological, mechanical, and engineering sciences where novel formulation and characterization techniques and methods are developed to probe, manipulate, control and

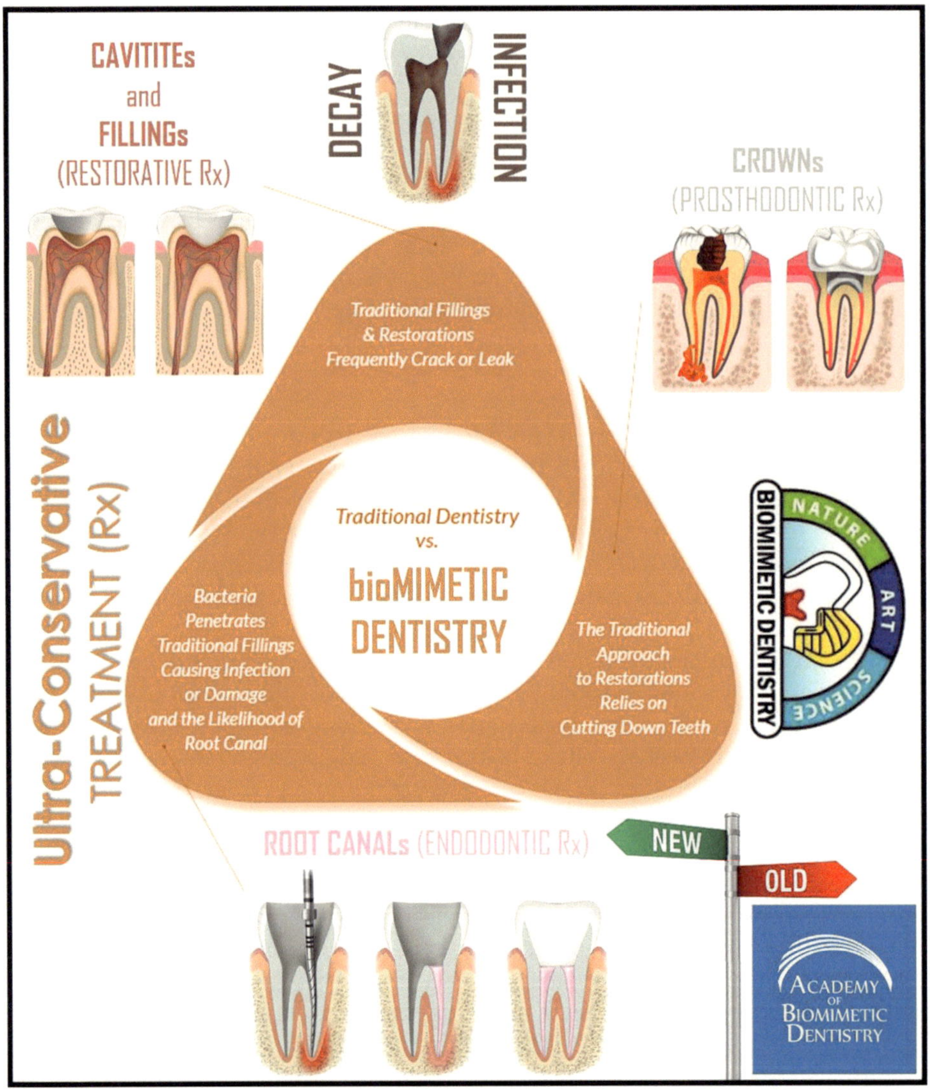

Figure 7.
Traditional versus bioMIMETIC and Ultra-Conservative Dentistry of today, with applications in dento-alveolar tissue restoration, replacement, and repair.

monitor single atoms and molecules (and combinations thereof). A nanoparticle (10^{-9} m) can be simply defined as a *small*, minute or tiny object that behaves as a whole unit in terms of its properties, characteristics, transport, and functions. Herein, it is perhaps worth mentioning that, today, the art, science, and engineering of nano-Systems are one of the most challenging and fastest-growing areas of nanotechnology, in general, and nano-biotechnology, in particular [28, 31, 32]. *Why nano-scale and nanoparticles per se?* Well, to simplify, as the size of the system decreases (from macro to micro to nano), a number of physical (amongst others) phenomena becomes more prominent and come into play. To rephrase further, an increase in the surface area to volume ratio leads to an enhancement of the behavior of atoms (more so on the surface of the nanoparticle rather than within/inside, herein), thus, altering the physical, chemical, mechanical, biological, optical, thermal, and catalytic properties of the whole material, or nano-System [28, 31, 32]. SPR (Surface Plasmon Resonance) and para-magnetism in metal-based magnetic nanoparticles, is a fine example. Anti-microbial property of Copper- or Silver-based nanoparticles [28, 38] is another, with increasing interest for utility in surgery and dentistry, given the superior inhibitory effect (sensitive as well resistant isolates of bacteria) against several microorganisms, including *Streptococcus mutans*, one of the main causative bacteria for Dental Caries or Decay. An illustration of biomimetics in dentistry versus traditional dentistry and its contribution to ultra-conservatism in the therapeutic strategies is shown in **Figure 7**.

5. bioMIMETIC cranio-facial tissue engineering and nanoDENTISTRY

The mechanical behavior of the cranio-maxillo-facial skeleton under physiological loads is among the least understood in the field of musculoskeletal bio-mechanics [38–40]. The musculoskeletal system contains a variety of supporting tissues, including muscle, bone, ligament, cartilage, tendon, and meniscus, which support the shape and structure of the body. Whether due to trauma, injuries, cancer, and/or diseases, defected, damaged, or lost tissue needs repair or replacement with healthy tissue [38–42]. Depending on the type of damaged tissue (whether it be cartilage, bone, skeletal muscle, tendon, and/or ligament), an extensive range of natural and synthetic (and composites thereof) biomaterials are available and possible, alongside simple and/or complex fabrication and formulation techniques, allowing to create a particular scaffold (porous scaffolds, 3D printed scaffolds, *in situ*-forming gels, bio-ceramics, and/or bioactive glass, electro-spun nano-Fibers, hydroxyapatite/collagen, and/or injectable hydrogels, amongst others); with an essential role in the regeneration, restoration, reconstruction, and repair (or replacement) of the musculoskeletal system, bearing in mind its governing structural bio-mechanics [32, 33]. In regenerative medicine, several factors should be taken into account when designing a system for successful organ regeneration using tissue engineering principles, including: (i) resident or transplanted cells should differentiate into specific cell types within a biomimetic matrix; (ii) the biomimetic matrix should provide biological and mechanical support for cell growth and function; (iii) the biomimetic matrix should allow for growth factor/cytokine permeation and physiological signals/cues; and (iv) the biomimetic matrix should have high engraftment efficiency [28, 30–32, 38]. Hence, designing, developing, characterizing, evaluating, and/or testing (then fine-tuning or optimizing) systems that encompass all of the above *ideal characteristics* have proved challenging. For example, the most commonly used *in vitro* culture techniques, today, do not mimic all of the micro- and nano- environmental factors that direct cell differentiation into a developing organ.

NANOTECHNOLGY in medicine (reputed as nanoMedicine) or/and in dentistry (reputed as nanoDentistry) has opened new realms and provided novel solutions for such demanding characteristics, properties, and needs (to full the criteria for proper tissue engineering and regenerative medicine/dentistry) [43, 44]. Specifically, it has done so by developing desirable and superior materials to control the physical/structural, chemical, biological, and mechanical micro-environment, *critical for*, successful cell (gene- and drug-) delivery and tissue regeneration [28, 32–37]. Therefore, manipulating biomaterials to create material surfaces and structures with nano-scale features (nano-bioMaterials) not only can help mimic the native micro- and nano-environment of cells, but also can trigger (and direct or guide – control and modulate) select cell adhesion, growth, proliferation, and differentiation with/without the use of drugs [28, 45–48]. **Figure 8** depicts the role of natural polymers (polysaccharides) from different eco-sources in designing, formulating, and developing nano-bioMaterials for Tissue bioEngineering and Regenerative Medicine.

The generation of new organic/inorganic analogs [45–48] of Carbon-based nano-bioMaterials (including graphene) for stem-cell-based tissue engineering, using nano-fibrous scaffolds and hybrid hydrogels as carriers or matrices, is a fine example. Further, since bone is a nano-composite, by nature, containing nano-scale building blocks (mainly collagen fibrils and mineral hydroxyapatite plates), the use of biodegradable conductive nano-composites is attractive for orthopedic, cranio-maxillo-facial and oro-dental applications [31, 35–37, 39–41, 44, 49, 50]. Therefore, many examples are found in literature where nano-bioMaterials allow for better hard, bone (and soft) tissue regeneration, providing a better surface and physicochemical properties for osteoblast attachment and long-term function. Better mechanical properties for certain load-bearing conditions in orthopaedic applications [31, 36, 39–41, 44, 49, 50] as well as in oral and maxillofacial implantology (*i.e.*, dental Titanium or

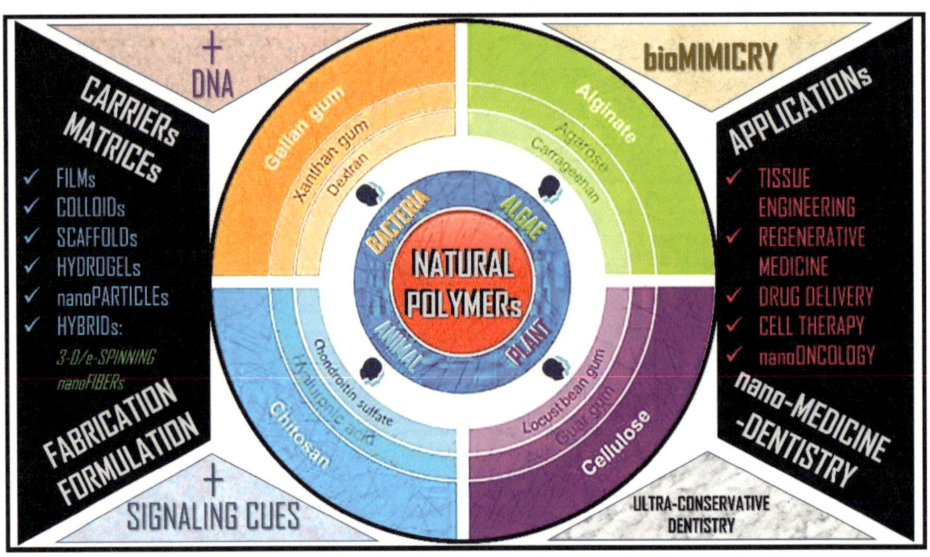

Figure 8.
Natural Polymer-based nanomedicine and nanoDENTISTRY. Natural Polysaccharides are abundant, biocompatible, biodegradable and have been incorporated, whether alone or in combination (in addition to other cells, genes, drugs, molecules, etc.) in carriers, matrices, scaffolds, and delivery systems for cell therapy, pharmaceutics, oncology, tissue engineering, and regenerative applications.

Ti implants and Ti implant-retained dentures, also known as overdentures—*for the edentate or completely edentulous patient*) [51, 52] have been also feasible, herein.

To recap, scaffolds developed for tissue engineering and tissular regeneration, trying to *mimic* the natural extracellular matrix, require a good control of porosity, inter-connectivity between pores, and even the size of pores. The macro-, micro-, and nano-architecture of 3D scaffolds (for soft and hard tissue engineering, repair and regeneration) have a primordial relevance to replicate the structural complexity of living tissues. Today, hybrid systems considering multiple material mixtures and a combination of fabrication processes are fundamental to *mimic* the natural tissues by providing multi-phasic or multi-material structures, accommodate growth factors and cells, and supply the signaling cues, vital to guide cell adhesion and proliferation [31, 35–37, 39–41, 44, 49, 50]. Remember, depending on the nature of the target defective/damaged/lost tissue and the required mechanical, chemical, and biological properties, different biomaterials can be employed, either singly or in combination, or with other additive materials [31, 35–37, 39–41, 44, 49, 50]. Further, although bone, described above as a natural nano-composite, a connective tissue with the ability to heal/regenerate spontaneously, in several specific cases, such as critical-sized defects, it fails to do so. It is important to consider the mechanical conditions affecting bony regeneration and the influence of the surrounding environment [49, 53, 54]. The quality of the soft tissue covering the defect is detrimental, also [50, 54].

In craniofacial defects, for example, reconstructive surgeons collaborate with engineers to design and create, fabricate, or formulate complex 3D geometrically structured or printed scaffolds, for a more personalized (patient-specific) corrective approach, regardless of the size or anatomic location of the cranio-maxillo-facial defect [38, 50, 54]. The potential benefits of using a tissue engineering approach include reduced donor site morbidity, shortened operative time, decreased technical difficulty of the repair, and, most important, ability to closely *mimic* the *in vivo* micro-environment, to attempt recapitulating normal cranio-maxillo-facial development [38, 50, 54–59]. Whether resulting from trauma, pathology, and/or osteonecrosis, mechanically stable and space-maintaining scaffolds tailored to site-specific defects, should be osteogenic, osteoconductive, osteoinductive while promoting angiogenesis and vasculogenesis [56, 59, 60]. Herein, it is critical to achieve a balance between the engineering the physiological criteria, for more safe and efficacious *devices*, to prevent and overcome infection, wound dehiscence, pre-mature resorption, and/or graft/implant failure. As one of the three essential components of tissular regenerative engineering, biomimetic NANOTECHNOLOGY is shifting the paradigm of employable nano-bioMaterials and holds promise for the future generation of substitute tissue grafts, including 3D printed and hybrid scaffolds [60].

REGENERATIVE DENTISTRY [61, 62] approaches include the formation of new enamel, dentin, pulp, periodontal ligament, and alveolar bone after tooth damage due to genetic pathology, traumatic injuries, caries, and periodontal lesions. Recent progress in the fields of mesenchymal stem/stromal cell biology (including, oral- and dental-derived), tissue engineering, and nanotechnology is considerable and offers new concepts and promising opportunities to repair, restore, and/or replace damaged or missing oral and dental tissues; *modern dental treatment strategies and protocols*. Likewise, scaffolds are *key* elements for the success of oro-dental tissue regeneration [12, 39, 61–63]. Indeed, in regenerative dentistry (dento-alveolar structures and tissues), emerging technologies such as bio-printing continue to be developed to solve some of the shortcomings of traditional tissue engineering approaches [56, 61]. For example, automated bio-printing facilitates the creation of geometrical patient-specific scaffold designs and in parallel permits the precise placement of cells and biological molecules thereby producing a biological or bioactive personalized scaffold [56, 61, 64, 65]. Yet, translation into a clinical product and our patients is still challenging, as the choice of a suitable material to encapsulate cells (and the other essential biological factors) in the development of the "bio-inks," for instance, for ultimately or pertinently bio-printing the *heterogeneous* dento-alveolar tissues (such as the pulp-dentin complex and the periodontal ligament, surrounding the tooth and filling the space between the tooth and its socket in the alveolar bone), is intricate and still posing technical hurdles and economic constraints [61, 64–67]. Yet, it is perhaps worth mentioning that R&D&I [60–69] on bio-printing dento-alveolar tissue, in particular, is still in the early stages, however, alongside accruing advancements in nano-science (novel sophisticated composites and fillers), tissue engineering techniques, biomimetic scaffolds (that also provide *geometric* cues for *de novo* tissue regeneration), stem/stromal cell-based therapeutics, autologous cell grafting, platelet concentrates (including platelet-rich fibrin), inkjet-based 3D printing (fabricated calcium phosphate or CaP scaffolds, for example), additive manufacturing (by extrusion: deformation + solidification, laser-assisted sintering or polymerization, for example), and rapid prototyping (for visual and informative models) , laser-scanning, computer-aided design and computer-aided manufacturing (that can manipulate 3D computed tomography images of bone, for example, virtually and in real-time) as well as in the available imaging and diagnostic techniques (that can verify and quantify the degree of mineralization and vascularization of the implanted scaffolds at the earliest stages of tissue regeneration), to list a few, the future, undoubtedly, brings new horizons and ample hope for innovative (and a new era of) oro-dental and cranio-maxillo-facial treatments, to all our patients and clinics.

6. Conclusions

BioMIMETICs is a research, development, and innovation field that is attaining growing interest and accruing prominence through an unprecedented flood of new discoveries and technologies in biology and creative engineering. From less than a hundred articles per year in the 1990s to several thousand studies and publications, annually, in the last decade, with impact "theoretical" across disciplines and diverse biomedical and bioengineering themes, including artificial intelligence, robotics, bioinformatics and omics. Biomimetics, today, is the leading scientifically-relevant paradigm for the innovative *potential* design and guiding the advancement of new methods and devices, for a high scientific, medical, and socio-economic impact, soon.

REGENERATIVE DENTISTRY employs the accruing understanding of cellular and molecular biology into the creative (and patient-specific) design of oral, dental, and cranio-maxillo-facial therapies that ultimately aim to restore, repair, rejuvenate, replace, and/or regenerate defected, damaged/injured or lost tissues. For the future of clinical and surgical dentistry, the grand challenge for bioMIMICRY incorporating and influencing (or inspiring) tissue engineering and regenerative nano-bioMaterials is to boost funding, expedite high-quality research, clinical translation, and education.

DRUG DELIVERY, in particular release-controlled and -targeted systems, of therapeutic agents (including cells, proteins, genes, anti-microbials, and so on), using bio-polymeric materials (such as, injectable and/or stimuli-responsive hydrogels and nano-Fibers whether electro-spun, 3D printed or produced through a combinatorial/

Hybrid approach/methodology) and colloidal vesicles (such as, liposomes, solid lipid nanoparticles, core-shell nanocapsules, and micelles), have been introduced, yet, *critical* challenges still remain, as most of technologies are *subpar* and fail to reach clinical expectations. Novel pharmaco-kinetic approaches, inspired by nature, could perhaps provide alternative, superior, and safer dose-responsive, and targeted solutions, for more precise and sustainable pharmaceutics.

Bio-TOOTH [70], the bio-Engineered and fully functional tooth *explant* from adult (embryonic-like) cells, as a replacement for the Ti-metal dental implant, although not yet realized, it is becoming more realistic, with the recent advances discussed in this chapter. Perhaps, to better understand on how to realize the bio-tooth, someday, it is good to seek inspiration and clues from nature and the evolution studies on species, such as fish and reptiles, which continuously replace their teeth? BioMIMICRY *may* help.

Acknowledgements

This work was supported by operating grants provided to the BioMAT'X I+D+I Group (HAiDAR R&D&I LAB: Laboratorio de Biomateriales, Farmacéuticos y Bioingeniería de Tejidos Cráneo Máxilo-Facial), a member of the Centro de Investigación e Innovación Biomédica (CiiB), Faculty of Medicine, Universidad de los Andes, through the awarded project funds: (1) NAM-USA / ANID-Chile # NAM 21I0022 "SockGEL/PLUG" (2021–2023), (2) CORFO Crea y Valida # 21CVC2-183649 (2022–2024) "bioFLOSS", (3) CORFO Crea y Valida—Proyecto de I+D+i Colaborativo—Reactívate # 22CVC2-218196 (2022–2024) "EndoCOAT'X," and FONDEF IDEA DE I+D, SUBDIRECCIÓN DE INVESTIGACIÓN APLICADA/ANID 2022, # ID22I10215 (2023–2025) "maxSALIVA."

Conflict of interest

The author declares no conflict of interest.

Notes/thanks/other declarations

I would like to thank Ms. Ana Javor and Mr. Josip Knapic, Author Service Manager(s) at IntechOpen for their constant support, communication, feedback and in the timely preparation of our project "Biomimetics – Bridging the Gap," with content, to the best of abilities, *different* from what has been previously published. THANK YOU and hope for the next.

R&D&I FOCUS and INTEREST:
The End-User (Clinician and Patient)-oriented design, development, characterization, evaluation, fine-tuning and translation of bio-nano-technology; biopolymers; bioceramics, pharmaceutical delivery systems and medical devices for the repair; restoration, reconstruction and regeneration of challenging cranio-maxillo-facial, oro-dental, orthopaedic defects, and R_x / management of cardiovascular and oncologic conditions.
Prof. Dr. ZS HAIDAR
PROFESSOR of BioENGINEERING and FOUNDER/DIRECTOR of BioMAT'X R&D&I Chile

Author details

Ziyad S. Haidar
BioMAT'X (HAiDAR LAB) R&D&I, Biomedical Research and Innovation Center (CiiB), Faculties of Medicine and Dentistry, University of the Andes, Santiago, Chile

*Address all correspondence to: zhaidar@uandes.cl; zhaidar78@gmail.com

References

[1] Hargroves K, Smith M. Innovation inspired by nature: Biomimicry. ECOS. 2006;**2006**(129):27-29

[2] Shimomura M. The new trends in next generation biomimetics material technology: Learning from biodiversity. Science and Technology. 2010;**37**:53-75

[3] Vincent JF, Bogatyreva OA, Bogatyrev NR, Bowyer A, Pahl A-K. Biomimetics: Its practice and theory. Journal of Royal Society Interface. 2006;**3**(9):471-482

[4] Freedman R. The Wright Brothers: How They Invented the Airplane. New York: Holiday House; 1991

[5] Jakab PL. Visions of a Flying Machine: The Wright Brothers and the Process of Invention. Washington, DC, USA: Smithsonian Institution Press; 1990

[6] Rinaldi A. Naturally better. Science and technology are looking to nature's successful designs for inspiration. EMBO Report. 2007;**8**(11):995

[7] Pianka ER, Sweet SS. Integrative biology of sticky feet in geckos. BioEssays. 2005;**27**(6):647-652

[8] Hwang J, Jeong Y, Park JM, Lee KH, Hong JW, Choi J. Biomimetics: Forecasting the future of science, engineering, and medicine. International Journal of Nanomedicine. 2015;**10**:5701-5713

[9] Srinivasan K, Chitra S. Emerging trends in oral health profession: The biomimetic – A review. Archives of Dental Medical Research. 2015;**1**:40-47

[10] Ikeda E, Morita R, Nakao K, Ishida K, Nakamura T, Takano-Yamamoto T, et al. Fully functional bioengineered tooth replacement as an organ replacement therapy. Proceedings of the National Academy Science USA. 2009;**106**:13475-13480

[11] Gandolfi MG, Taddei P, Siboni F, Modena E, De Stefano ED, Prati C. Biomimetic remineralization of human dentin using promising innovative calcium-silicate hybrid "smart" materials. Dental Materials. 2011;**27**:1055-1069

[12] Upadhyay A, Pillai S, Khayambashi P, et al. Biomimetic aspects of oral and dentofacial regeneration. Biomimetics (Basel). 2020;**5**(4):51

[13] Gong T, Heng BC, Lo ECM, Zhang C. Current advance and future prospects of tissue engineering approach to dentin/pulp regenerative therapy. Stem Cells International. 2016;**2016**:9204574

[14] Benyus JM. Biomimicry: Innovation Inspired by Nature. New York: William Morrow and Co.; 1997. p. 308

[15] Benyus JM. Biomimicry. HarperCollins e-books: New York, USA; 2009

[16] Dunlap RA. The Golden Ratio and Fibonacci Numbers. Singapore: World Scientific; 1988

[17] Livio M. The Golden Ratio: The Story of phi, the World's Most Astonishing Number. Broadway Books. Portland, OR, USA; 2008

[18] Prusinkiewicz P, Lindenmayer A. The algorithmic Beauty of Plants. New York, USA: Springer Science & Business Media; 2012

[19] Aram MR. Aesthetics of artistic expression in Qur'an. Burhan Journal of Qur'anic Studies. 2016;**1**(1):1-19

[20] Ilieva L, Ursano I, Traista L, Hoffmann B, Dahy H. Biomimicry as a sustainable design methodology—Introducing the 'Biomimicry for Sustainability' framework. Biomimetics. 2022;**7**(2):37

[21] Perera AS, Coppens MO. Re-designing materials for biomedical applications: From biomimicry to nature-inspired chemical engineering. Philosophical Transactions: A Mathematical, Physical and Engineering Science. 2019;**377**(2138):201

[22] Yang SY et al. A bio-inspired swellable microneedle adhesive for mechanical interlocking with tissue. Nature Communication. 2013;**4**:1702

[23] https://www.medgadget.com/2015/11/hold-tu-delft-developing-surgical-tools-snaking-body.html

[24] Cui Y, Gong H, Wang Y, Li D, Bai H. A thermally insulating textile inspired by polar bear hair. Advanced Materials. 2018;**30**(14):e1706807

[25] Luo J, Guo P, Lin Y. A novel modular biomimetic live working robot for power distribution line. Machines. 2022;**10**(3):195

[26] Niederberger C, Zola Z. Medicine and architecture: A blueprint. Fertility and Sterility. 2018;**109**(4):592-593

[27] Amini AR, Laurencin CT, Nukavarapu SP. Bone tissue engineering: Recent advances and challenges. Critical Reviews in Biomedical Engineering. 2012;**40**:363-408

[28] Haidar ZS. Bio-inspired/-functional colloidal core-shell polymeric-based nanosystems: Technology promise in tissue engineering, bioimaging and nanomedicine. Polymers. 2010;**2**:323-352

[29] Cao L, Wang J, Hou J, Xing W, Liu C. Vascularization and bone regeneration in a critical sized defect using 2-N,6-O-sulfated chitosan nanoparticles incorporating BMP-2. Biomaterials. 2014;**35**:684-698

[30] Hubbell JA. Materials as morphogenetic guides in tissue engineering. Current Opinion in Biotechnology. 2003;**5**:551-558

[31] Haidar ZS, Hamdy RC, Tabrizian M. Delivery of recombinant bone morphogenetic proteins for bone regeneration and repair. Part A: Current challenges in BMP delivery. Biotechnology Letter. 2009;**31**:1817-1824

[32] Haidar ZS, Hamdy RC, Tabrizian M. Delivery of recombinant bone morphogenetic proteins for bone regeneration and repair. Part B: Delivery systems for BMPs in orthopaedic and craniofacial tissue engineering. Biotechnology Letters. 2009;**31**:1825-1835

[33] Joo V, Ramasamy T, Haidar ZS. A novel self-assembled liposome-based polymeric hydrogel for cranio-maxillofacial applications: Preliminary findings. Polymers. 2011;**3**(2):967-974

[34] Haidar ZS, Hamdy RC, Tabrizian M. Protein release kinetics for core-shell hybrid nanoparticles based on the layer-by-layer assembly of alginate and chitosan on liposomes. Biomaterials. 2008;**29**:1207-1215

[35] Haidar ZS, Azari F, Hamdy RC, Tabrizian M. Modulated release of OP-1 and enhanced pre-osteoblast differentiation using a core-shell nanoparticulate system. Journal of

Biomedical Materials Research. Part A. 2009;**91**:919-928

[36] Haidar ZS, Tabrizian M, Hamdy RC. A hybrid OP-1 delivery system enhances new bone regeneration and consolidation in a rabbit model of distraction osteogenesis. Growth Factors. 2010;**28**:44-55

[37] Haidar ZS, Hamdy RC, Tabrizian M. Biocompatibility and safety of a hybrid core-shell nanoparticulate OP-1 delivery system intramuscularly administered in rats. Biomaterials. 2010;**31**:2746-2754

[38] Tevlin R, McArdle A, Atashroo D, et al. Biomaterials for craniofacial bone engineering. Journal of Dental Research. 2014;**93**(12):1187-1195

[39] Zumarán CC, Parra MV, Olate SA, Fernández EG, Muñoz FT, Haidar ZS. The 3 R's for platelet-rich fibrin: A "Super" tri-dimensional biomaterial for contemporary naturally-guided oro-maxillo-facial soft and hard tissue repair, reconstruction and regeneration. Materials (Basel). 2018;**11**(8):1293

[40] Muñoz F, Jiménez C, Espinoza D, Vervelle A, Beugnet J, Haidar Z. Use of leukocyte and platelet-rich fibrin (L-PRF) in periodontally accelerated osteogenic orthodontics (PAOO): Clinical effects on edema and pain. Journal of Clinical and Experimental Dentistry. 2016;**8**:e119-e124

[41] Muñoz F, Haidar Z. L-PRF for use in oro-maxillo-facial surgeries: What do we know? Journal of Oral Research. 2018;7:88-90

[42] Pripatnanont P, Nuntanaranont T, Vongvatcharanon S, Phurisat K. The primacy of platelet-rich fibrin on bone regeneration of various grafts in rabbit's calvarial defects. Journal of Cranio-Maxillo-Facial Surgery. 2013;**41**:e191-e200

[43] Haidar ZS. NanoDentistry: Perspectives on the role of NanoBiotechnology in biomaterials. Pharmaceutics and BioDental Tissue Engineering, EC Dental Science. 2015;**3**:506-507

[44] Haidar ZS. nanoBONE: Re-visiting Osseo-Reconstruction and -Repair … with a nanoTwist. Journal of Oral Research. 2015;**10**:1-6

[45] Urrejola MC, Soto LV, Zumarán CC, Peñaloza JP, Álvarez B, Fuentevilla I, et al. Sistemas de Nanopartículas Poliméricas I: de Biodetección y Monitoreo de Glucosa en Diabetes a Bioimagen, Nano-Oncología, Terapia Génica, Ingeniería de Tejidos / Regeneración a Nano-Odontología. International Journal of Morphology. 2018;**36**(4):1490-1499

[46] Urrejola MC, Soto LV, Zumarán CC, Peñaloza JP, Álvarez B, Fuentevilla I, et al. Sistemas de Nanopartículas Poliméricas II: Estructura, Métodos de Elaboración, Características, Propiedades, Biofuncionalización y Tecnologías de Auto-Ensamblaje Capa por Capa (Layer-by-Layer Self-Assembly). International Journal of Morphology. 2018;**36**(4):1463-1471

[47] Dissanayaka WL, Zhang C. Scaffold-based and Scaffold-free strategies in dental pulp regeneration. Journal of Endodontia. 2020;**46**(9S):S81-S89

[48] Bar JK, Kowalczyk T, Grelewski PG, Stamnitz S, Paprocka M, Lis J, et al. Characterization of biological properties of dental pulp stem cells grown on an electrospun poly(l-lactide-co-caprolactone) scaffold. Materials. 2022;**15**(5):1900

[49] Olate S et al. NanoBioTechnology-guided distraction osteogenesis and

histiogenesis. Journal of Oral Research. 2017;**6**(6):142-144

[50] Olate S et al. Growth factor-assisted distraction osteogenesis and histiogenesis. Journal of Oral Research. 2017;**6**(5):112-114

[51] Haidar ZS, Silva RB. Ex-vivo thermo-dynamic conductivity model for osseointegrated titanium fixtures. Journal of Oral Research. 2020;**34**:38

[52] Samara SA, Haidar ZS. Re-visiting Edentulism: Complete removable dentures vs implant overdentures. Journal of Oral Health & Community Dentistry. 2011;**5**:107-109

[53] Maître G, Farias D, et al. Horizontal ridge augmentation of a single atrophic site in the anterior maxilla using hydroxyapatite and *rh*BMP-2. International Journal of Morphology. 2020;**38**(5):1426-1433

[54] Ravelo V, Olate G, Huentequeo-Molina C, Haidar ZS, Parra M, Vásquez B, et al. Morfología Ósea Facial en Cirugía Ortognática. ¿Existe. Tendencia Hacia el Avance Facial? International Journal of Morphology. 2021;**39**(4):1116-1122

[55] Haidar ZS. 3D printed titanium implants: colossal FDA-approved leap towards personalized. Journal of Oral Research. 2017;**6**(11):282-284

[56] Shen C, Witek L, Flores RL, Tovar N, Torroni A, Coelho PG, et al. Three-dimensional printing for craniofacial bone tissue engineering. Tissue Engineering. Part A. 2020;**26**(23-24):1303-1311

[57] Nguyen C, Young S, Kretlow JD, et al. Surface characteristics of biomaterials used for space maintenance in a mandibular defect: A pilot animal study. Journal of Oral and Maxillofacial Surgery. 2011;**69**:11

[58] Smith BT, Shum J, Wong M, et al. Bone tissue engineering challenges in oral & maxillofacial surgery. Advances in Experimental Medicine and Biology. 2015;**881**:57

[59] Bharadwaz A, Jayasuriya AC. Recent trends in the application of widely used natural and synthetic polymer nanocomposites in bone tissue regeneration. Materials Science & Engineering. C, Materials for Biological Applications. 2020;**110**:110698

[60] Yilmaz B, Tahmasebifar A, Baran ET. Bioprinting technologies in tissue engineering. Advances in Biochemical Engineering/Biotechnology. 2020;**171**:279

[61] Sharpe P. Regenerative dentistry. Frontiers in Dental Medicine. 2020;**1**:3

[62] Haidar ZS. Regenerative Endodontics and the promise beyond dental pulp disease repair. Journal of Oral Research. 2018;7(2):49

[63] Damsaz M, Castagnoli CZ, EshghpourM, AlamdariDH, AlamdariAH, Noujeim ZEF, et al. Evidence-based clinical efficacy of leukocyte and platelet-rich fibrin in maxillary sinus floor lift, graft and surgical augmentation procedures. Frontier in Surgery. 2020;**7**:537138

[64] Ong CS, Yesantharao P, Huang CY, et al. 3D bioprinting using stem cells. Pediatric Research. 2018;**83**:223

[65] Bekisz JM, Flores RL, Witek L, et al. Dipyridamole enhances osteogenesis of three-dimensionally printed bioactive ceramic scaffolds in calvarial defects. Journal of Cranio-Maxillo-Facial Surgery. 2018;**46**:237

[66] Zein I, Hutmacher DW, Tan KC, et al. Fused deposition modeling of novel scaffold architectures for tissue engineering applications. Biomaterials. 2002;**23**:1169

[67] Silva NR, Witek L, Coelho PG, et al. Additive CAD/CAM process for dental prostheses. Journal of Prosthodontics. 2011;**20**:93

[68] Fernandes HR, Gaddam A, Rebelo A, et al. Bioactive glasses and glass-ceramics for healthcare applications in bone regeneration and tissue engineering. Materials (Basel, Switzerland). 2018;**11**:2530

[69] Ashammakhi N, Kaarela O. Three-dimensional bioprinting can help bone. The Journal of Craniofacial Surgery. 2018;**29**:9

[70] Mitsiadis TA, Harada H. Regenerated teeth: The future of tooth replacement. Regenerative Medicine. 2015;**10**:5-8

Section 2

Biomimetic Tissue Engineering

Chapter 3

Perspective Chapter: Design and Characterization of Natural and Synthetic Soft Polymeric Materials with Biomimetic 3D Microarchitecture for Tissue Engineering and Medical Applications

Ching-Cheng Huang and Masashi Shiotsuki

Abstract

Continuous work and developments in biomedical materials used in three-dimensional (3D) bioprinting have contributed to significant growth of 3D bioprinting applications in the production of personalized tissue-repairing membrane, skin graft, prostheses, medication delivery system, and 3D tissue engineering and regenerative medicine scaffolds. The design of clinic products and devices focus on new natural and synthetic biomedical materials employed for therapeutic applications in different 3D bioprinting technologies. Design and characterization of natural and synthetic soft polymeric materials with biomimetic 3D microarchitecture were considered. The natural soft polymeric materials would focus on new design bioinspired membranes containing supercritical fluids-decellularized dermal scaffolds for 3D bioprinting potential applications. Synthetic soft polymeric materials would focus on bioinspired polyvinyl alcohol (b-PVA) matrix with structural foam-wall microarchitectures. Characterization, thermal stability, and cell morphology of the b-PVA and the corresponding collagen-modified b-PVA were employed to evaluate their potential tissue engineering applications. Also, the b-PVA materials were conductive to HepG2 cells proliferation, migration, and expression, which might serve as a promising liver cell culture carrier to be used in the biological artificial liver reactor. TGA, DTG, DSC, SEM, and FTIR were employed to build up the effective system identification approach for biomimetic structure, stability, purity, and safety of target soft matrix.

Keywords: biomimetic, three-dimensional, microarchitecture, design-thinking, supercritical fluids

1. Introduction

Bioprinting, a type of three-dimensional (3D) bioprinting, uses cells and other biological materials as "inks" to fabricate 3D biological structures. Bioprinted materials have the potential to repair damaged organs, cells, and tissues in the human body. Skin, bone, and blood vessels may be bioprinted. Although a variety of tissue engineering strategies combining cells and biomaterial scaffolds have been investigated for human tissues regeneration (e.g., using natural biomaterials and/or synthetic polymers [1]), the success of this approach still strongly depends on the development of more suitable scaffolds. In this sense, one of the main challenges is to accurately reproduce the complex 3D anatomy with personalized shape and size [2–4]. Efforts to address these issues have led to an increased interest in the application of three-dimensional (3D) bioprinting in the tissue engineering field. 3D bioprinting is a biofabrication method that uses computer-aided design (CAD) and additive layer manufacturing technique to precisely deposit bioinks (basically comprising a mixture of biomaterials with cells, with or without bioactive molecules) in a predesigned manner to create 3D bioengineered living structures and to generate artificial tissue and organs [5, 6]. Using this novel printing technology, biomaterials were possible to fabricate a patient-specific scaffold reproducing individual shape, size, and macrostructure of the native tissue [7, 8]. An implant's effectiveness depends upon the form of biomaterial used in its manufacture. A suitable material for implants should be biocompatible, sterile, mechanically stable, and simple to shape [9]. 3D printing technologies have been breaking new ground in the medical industries in order to build patient-specific devices embedded in bioactive drugs, cells, and proteins. 3D printing was a non-exclusive concept that defines various techniques for layer-by-layer construction. 3D bioprinting has progressed the method of printing standard biocompatible materials and even actual cells into difficult 3D dimensional tissue buildings [10], with the ability to create ideal tissues and organs appropriate for different biomedical uses, such as organ transplantation [11]. In recent years, many bioprinting techniques have been developed to store cells and hydrogels together, altered and used as cell printers to print polymers [12]. In such printers, cell suspensions or cell totals are mounted on printer extruder, and the printing mechanism is controlled. Continuous work and developments in biomaterials used in 3D printing have contributed to significant growth of 3D printing applications in the production of personalized tissue-repairing membrane, skin graft, prostheses, medication delivery system, and 3D tissue engineering and regenerative medicine scaffolds. The design of clinic products and devices focus on new natural and synthetic biomedical materials used for therapeutic applications in different 3D bioprinting technologies. Many specific forms of medical 3D bioprinting technology are explored in depth, including extrusion-based bioprinting and inkjet printing processes, the specific therapeutic uses, various types of biomaterial used today, and the major shortcomings are being studied in depth.

Bioinspired structure-guided tissue/cell in vivo growth or vitro culturing tissue/cell in vivo growth and in vitro culturing are remarkably important for implantable medical devices and drug evaluation [13]. To achieve desired growth and reproduction, it is necessary to construct the culturing conditions and the growing structure as close to living being as possible. Bioinspired surface was applied to the implants to improve the implanting quality. Bioinspired porous structure was built on the scaffolds to guide and improve the tissue growth.

2. Design and preparation of natural soft polymeric materials with biomimetic 3D microarchitecture containing decellularized extracellular matrix

Tree-dimensional (3D) bioprinting showed potential in tissue engineering and regenerative applications due to its overwhelming advantages over other approaches. In order to promote the functions of bioprinted tissues, the development of novel and versatile bioinks will have crucial implications [14]. Natural materials were famous for the excellent biocompatibility and abundance, among which sodium alginate mixed with gelatin has been widely used as bioink for extrusion-based 3D bioprinting [15]. Despite advance in bioprinting and bio-fabrication during the past decade, fabricating complex and functional tissue constructs that mimic their natural counterparts still remains a challenge. Bioink optimization is considered as one of main challenges in cell-laden 3D bioprinting [16].

Numerous materials have been proposed, modified, and employed for medical bioprinting applications such as scaffolds for skin and bone tissue reconstruction such as synthetic materials and natural materials [17–29]. In bioprinting applications, biomedical materials could be extruded through a print head either by pneumatic pressure or mechanical force. Since the process did not involve any heating procedures, it was most commonly used for preparing tissue engineering constructs with cells and growth hormones laden. Bioinks were the biomedical materials laden with cells and other biological materials and used for 3D bioprinting. The 3D bioprinting process allowed for the deposition of small units of cells accurately, with minimal process-induced cell damage. Advantages such as precise deposition of cells control over the rate of cell distribution and process speed had greatly increased the applications of this technology in fabricating living scaffolds. A wide range of materials with varied viscosities and high cell density aggregates could be 3D printed using this technique. A large variety of polymers were under research for the use in bioprinting technology [18]. Natural polymers, including collagen [19], gelatin [20], decellularized extracellular matrix(dECM) [21], alginate [22], chitosan [23], cellulose [24], agarose [25], silk protein [26], and hyaluronic acid (HA) [27, 28], and photopolymerizable gelatin and hyaluronic acid [29] were commonly used in bioinks for 3D bioprinting (**Figure 1**). Often these bioinks are post-processed either by chemical or UV cross-linking to enhance the constructs' mechanical properties. Depending on the type of polymer used in the bioink, biological tissues and scaffolds of varied complexity can be fabricated.

Decellularized extracellular matrix (dECM) scaffolds had a lot of collagens, which constitute the main structural element of the dECM, provide tensile strength, regulate cell adhesion, support migration, and direct tissue development. Dense connective tissue is an abundant source of dECM scaffolds, which can be prepared and purified by a defatting and decellularizing procedure [21, 30]. The treatments combined with supercritical carbon dioxide and specific enzymes could be employed to prepare dECM scaffolds. Furthermore, a series of new biomedical composite materials containing dECM scaffolds and alginate could be designed and prepared for tissue engineering and bioprinting applications. The composite materials containing dECM scaffolds for biomimetic bioinks could be characterized by Fourier transform infrared spectroscopy (FTIR), thermo-gravimetric analysis (TGA), and scanning electron microscope (SEM) to get the results of identifications, thermal stabilities, and microstructures.

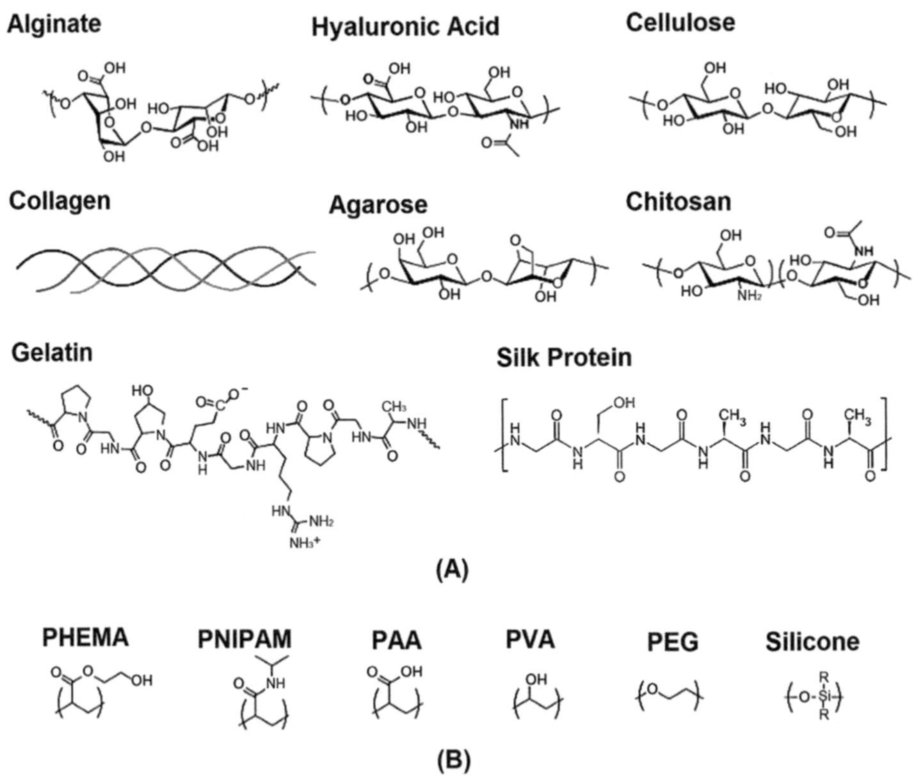

Figure 1.
Chemical structures of natural and synthetic soft polymeric materials with biomimetic 3D microarchitecture for tissue nanogineering, bioprinting, and medical applications. A) Natural soft polymeric material. B) Synthetic soft polymeric material.

2.1 Natural soft polymeric materials with biomimetic 3D microarchitecture containing decellularized extracellular matrix via supercritical fluid treatments

Decellularized extracellular matrix could be an important natural soft polymeric material with biomimetic 3D microarchitecture for bioprinting applications. In previous study, supercritical fluid of carbon dioxide ($ScCO_2$) was employed for preparation of designed decellularized extracellular matrix scaffolds for bioprinting applications. The $ScCO_2$ could be employed before or after decellularization treatments for effectively removing most fatty acids and tissues [21]. The $ScCO_2$ extraction could be performed for complete decellularization. Also, the steady thickness of about 0.5 mm of thinly sliced tissue sample, which could be obtained by using a designed tissue-cutting machine (Taiwan PARSD Pharm. Tech. Consulting Ltd Co.), was employed for complete decellularization.

Fourier transform infrared spectroscopy analysis was important for characterization of the natural soft polymeric material with biomimetic 3D microarchitecture containing decellularized extracellular matrix because of collagen segments, which could exhibit remarkable typical absorption bands of specific functional groups such as Amide I, Amide II, Amide III, Amide A, and Amide B could be an important for bioprinting applications. From the FTIR analysis of the original tissue such as porcine skin, absorption bands at 1452, 1400, 1337, 1240, 1203, and 1080 cm^{-1} were attributed to the amides III containing $\delta(CH_2)$, $\delta(CH_3)$, ν(C–N), and δ(N–H) absorptions of collagens in the original porcine skin. Amides I and amides II absorptions were found at 1632 and 1551 cm^{-1}, respectively. The absorption band at 3301 cm^{-1} δ(C–H) was attributed to the fatty acid of the original porcine skin. The absorption band at 1744 cm^{-1} δ(C=O) was attributed to the fatty acid. After supercritical carbon dioxide treatment and decellularization, the decellularized extracellular matrix could be formed and the absorption bands of fatty acids could not be observed in FTIR spectrum, demonstrating the effectiveness of the supercritical carbon dioxide treatment and the formation of natural soft polymeric materials. The resulting natural soft polymeric materials could be considered as a nano-bioscaffold. The microstructures of resulting membranes with dNBS-S were characterized by scanning electron microscope (SEM). Scanning electron micrographs of the decellularized samples such as decellularized porcine skin (dNBS-S), decellularized porcine liver(dNBS-L), decellularized porcine costal cartilages (dNBS-CC), and decellularized porcine elastic cartilage (dNBS-EC) after treatment with supercritical carbon dioxide are shown in **Figure 2(A)–(D)**, respectively. The pore space of the resulting natural soft polymeric materials from different tissues such as the skin, liver, costal cartilage, and elastic cartilage with diameters in a range of 10–250 μm was observed, which could be good nano-bioscaffolds for cell migration [21].

Further, L929 cells were cultured on the resulting natural soft polymeric material for a potential evaluation of good nano-bioscaffolds for tissue engineering. The morphology of L929 cells cultured on the resulting natural soft polymeric materials was also investigated by SEM (**Figure 3**). Significantly, most area was covered by L929

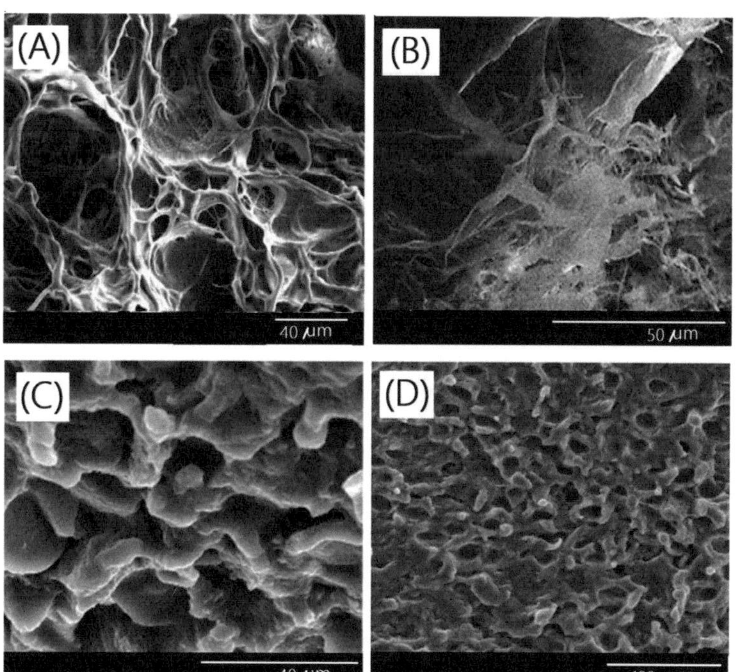

Figure 2.
Scanning electron micrographs of the samples: (A) decellularized porcine skin (dNBS-S), (B) decellularized porcine liver(dNBS-L), (C) decellularized porcine Costal cartilages(dNBS-CC), and (D) decellularized porcine elastic cartilage (dNBS-EC).

Figure 3.
SEM photograph of L929 cells grew on dNBS-S.

cells on the resulting natural soft polymeric materials such as dNBS-S after 3 days of culture was observed [21]. The cells that grew upon the dNBS-S were significantly observed. Perez-Puyana et al. also reported that the pore space of nano-bioscaffold was suggested to be small enough to establish a high specific surface area and large enough to allow cells to migrate into the microstructure (20–120 μm) [31].

2.2 Natural soft polymeric composite materials with biomimetic 3D microarchitecture containing alginate and decellularized extracellular matrix via supercritical fluid treatments

Decellularized extracellular matrix(dECM)/alginate composite materials with biomimetic 3D microarchitecture and cross-linked decellularized extracellular matrix/alginate composite materials with biomimetic 3D Microarchitecture could also be designed for tissue engineering and medical applications. Briefly, the desired amount of dECM powder was first dispersed completely in water. Then, alginate aqueous solution was homogenized thoroughly with the dispersed dECM solution. The alginate/dECM solutions were employed to form 3D printed membranes by using extrusion-based bioprinting, molding, and freeze-drying procedures for evaluation of new 3D bioprinting materials. Further, cross-linked decellularized extracellular matrix/alginate composite materials with biomimetic 3D Microarchitecture were designed and prepared by using ionic crosslinking procedure with $CaCl_2$ aqueous solution.

The incorporation of nano-bioscaffold(dNBS) in the alginate/decellularized extracellular matrix composite materials(AdNBS) with biomimetic 3D microarchitecture was done. The FTIR spectroscopy analysis was carried out to confirm the structures. In the spectrum of the AdNBS, the main absorption bands at around 1632 cm^{-1} (amide I, C-O, and C-N stretching), 1537 cm^{-1} (amide II) and 1242 cm^{-1} (amide III) were also observed. Because of the introduction of alginate segments, retaining the characteristic bands of pure sodium alginate was observed. The characteristic bands of AdNBS at around 1595 cm^{-1} (the carbonyl (C=O) bond) and 1408 cm^{-1} (asymmetric and symmetric stretching peaks of carboxylate salt groups) were visible [21].

It showed a stronger absorption band at 1595 cm^{-1} and two remarkably shoulders at 1632 and 1537 cm^{-1}, which were characteristic absorption of carbonyl groups of amide (amide I and amide II) of dNBS segments, which would confirm the formation of AdNBS composite materials remarkably. A higher absorption from 3600 to 3200 cm^{-1} would be observed in the spectrum, which suggested an increase of hydrogen bonds resulting from the interaction between dNBS molecule and alginate(ALG) molecule. The results of FTIR indicated the presence of dNBS in the composite materials as wells as the interaction between them. After cross-linking reaction, the cross-linked decellularized extracellular matrix/alginate composite materials with biomimetic 3D microarchitecture were obtained and the calcium alginate segments were formed. The -C-O-O-Ca-O-CO- structure made the C-O stretching vibration absorption increase and had an obvious absorption band at 1024 cm^{-1}, which indicated the formation of -CO-O-Ca-O-CO- structure in AdNBS materials [21].

The microstructure of resulting AdNBS composite materials was characterized by scanning electron microscopy (SEM). The micro-scaffold structure could be observed in the dNBS-S derived from porcine skin after treatment with supercritical carbon dioxide (**Figure 2(B)**) showed quite different morphology form ALG, which showed a micro-scaffold shape with narrow boundaries [21]. The averaged diameter of narrow boundaries was found in a range of 1–3 μm. The remarkable micro-scaffold structures were observed in AdNBS composite materials with various introduction ratios of dNBS-S and ALG. With the increasing introduction ratio of dNBS-S and ALG, the new combined micro-scaffold shapes of composite materials were observed with the spaces and smooth mixed boundaries. The averaged diameter of smooth mixed boundary was found in a range of 1–35 μm [21], which might provide relatively high structural and thermal stabilities.

Thermal stability of resulting natural soft polymeric materials with biomimetic 3D microarchitecture such as AdNBS-S could be evaluated by TGA (**Figure 4**). Thermogravimetric analysis of the ALG showed a maximum pyrolysis temperature (Tdmax) lower than 300 degrees. To enhance the thermal stability of ALG-based materials, the decellularized nano-bioscaffolds, dNBS-S, were introduced. The maximum pyrolysis temperature (Tdmax) of dNBS-S was higher than 300 degrees. The resulting AdNBS-S would be a new heat-resistant biomaterial for bioprinting applications. From TGA analysis of AdNBS-S, the main loss is presented in two different temperature ranges given by stage I(<150°C), stage II(150–250°C), and stage III (250–500°C). From TGA results of ALG, initial weight loss up to 150°C is found to be 15, 20, and 20 for non-cross-linked alginate(NCA), high cross-linked alginate(HCA) (5wt% $CaCl_{2(aq)}$), and relative low cross-linked alginate(LCA) (1wt% $CaCl_{2(aq)}$), respectively, due to the elimination of absorbed and bounded water molecules in the membrane. In case of cross-linked AdNBS-S (dNBS-S/ALG= 20/80), composite material with $CaCl_{2(aq)}$ such as relatively low-cross-linked AdNBS-S(dNBS-S/ALG=20/80) composite material(LCLAdNBS-S) (1wt% $CaCl_{2(aq)}$) and high-cross-linked AdNBS-S(dNBS-S/ALG=20/80) composite material(HCLAdNBS-S)(5wt% $CaCl_{2(aq)}$), weight loss would be increased comparing with the non-cross-linked AdNBS-S(dNBS-S/ALG=20/80) composite material sample (**Figures 4(C)**, **(F)**, and **(I)**). This increase may be due to more adsorption of water molecules present along with Ca^{2+} molecules while cross-linking with $CaCl_2$ aqueous solution. From TGA results, relatively high Tdmax(i) values of AdNBS-S composite materials were observed at 270–300°C in stage II comparing with the 250°C of non-cross-linked alginate(NCA) (**Figure 4(A)**). When the high concentration of $CaCl_2$ (5wt%) added, the relative high Tdmax(i) value of high-cross-linked alginate(HCA) was observed at ca. 270°C comparing with

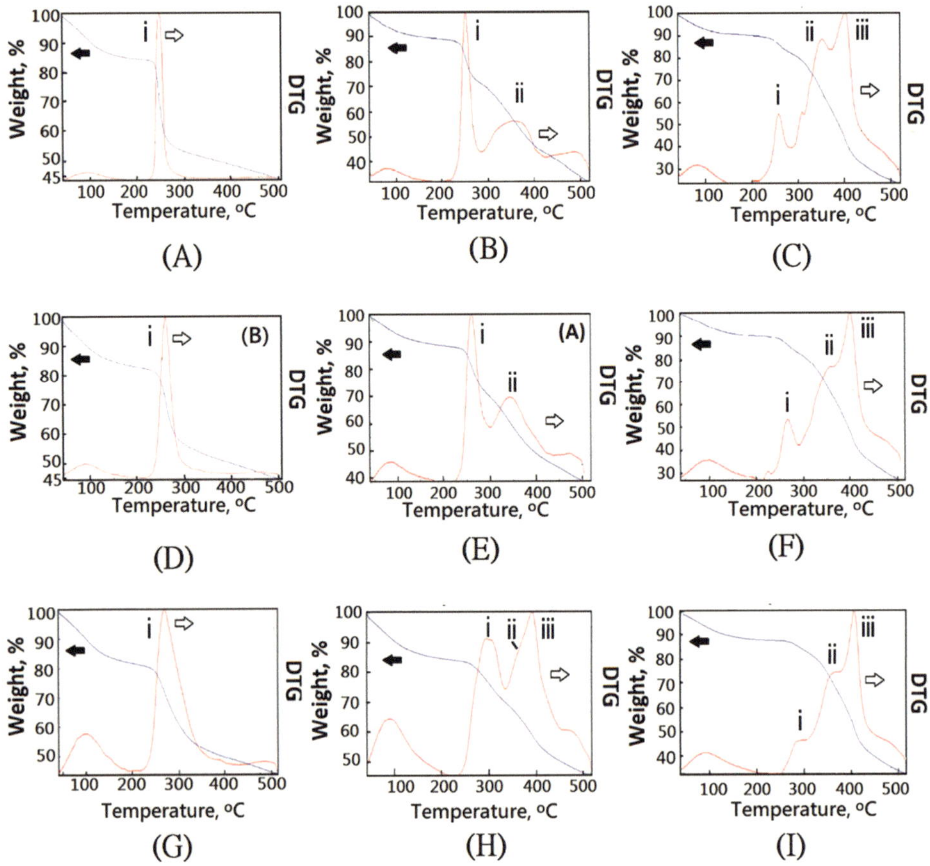

Figure 4.
Thermogravimetric analysis of the composite membranes with/without cross-linking reaction, (A) non-cross-linked alginate material (NCA); (B) non-cross-linked AdNBS-S (5/95) (NCHAdNBS-S); (C) non-cross-linked AdNBS-S (20/80) (NCLAdNBS-S); (D) relative low-cross-linked alginate(LCA) (1wt% $CaCl_{2(aq)}$; (E) relative low-cross-linked AdNBS-S (5/95) (LCHAdNBS-S) (1wt% $CaCl_{2(aq)}$); (F) relative low-cross-linked AdNBS-S (20/80) (LCLAdNBS-S) (1wt% $CaCl_{2(aq)}$); (G) high-cross-linked alginate(HCA) (5wt% $CaCl_{2(aq)}$); (H) high-cross-linked AdNBS-S (5/95)(HCHAdNBS-S) (5wt% $CaCl_{2(aq)}$); (I) high cross-linked AdNBS-S (20/80) (HCLAdNBS-S) (5wt% $CaCl_{2(aq)}$).

the 250°C of non-cross-linked alginate(NCA) and 252°C of relative low-cross-linked alginate molecule(LCA)(1wt%). Furthermore, when the high concentration of $CaCl_2$ (5wt%) was added, the relatively high Tdmax(i) values of high-cross-linked AdNBS-S such as HCHAdNBS-S and HCLAdNBS-S were observed at ca. 300°C comparing with the 270°C of high-cross-linked alginate (5wt%)(HCA) (**Figure 4(G)**). It would be due to the association between the ALG molecule and dNBS-S. Similarly, the relatively high Tdmax(ii) value of cross-linked AdNBS-S composite materials would be observed at ca.370°C. Particularly, much higher Tdmax(iii) values than Tdmax(ii) values of cross-linked AdNBS-S composite materials were observed at 400°C, which might be contributed to the formation of new mixed cross-linked network micro-structures of ALG and dNBS-S molecules.

When a little amount of dNBS-S was introduced into the AdNBS-S composite materials without $CaCl_2$, weak ionic association between –COOH group of alginate molecule and -NH_2 group of dNBS-S molecule was formed and which is difficult to build up the cross-linking structure. With an increasing introduction of dNBS-S to AdNBS-S composite materials, ordinary ionic association between –COOH group of alginate molecule and $-NH_2$ group of dNBS-S was employed to build weak ionic cross-linking microstructure. When a large amount of dNBS-S was introduced into the AdNBS-S composite materials without $CaCl_2$(NCLAdNBS-S), strong ionic association between -COOH group of ALG molecule and $-NH_2$ group of dNBS-S was employed to build up strong ionic cross-linking microstructure.

When a little amount of dNBS-S was introduced into the dNBS-S/alginate composite membrane with 1wt% $CaCl_2$(LCHAdNBS-S), weak ionic association between –COOH group of alginate molecule and $-NH_2$ group of dNBS-S and weak ionic associations among –COOH group of alginate molecule, Ca^{2+}, and –COOH group of alginate molecule could be found. The remarkable cross-linked microstructure was difficult to be observed. Tdmax(iii) value could not be found in DTG of LCHAdNBS-S results (Figure (E)). Weak ionic association between –COOH group of alginate molecule and $-NH_2$ group of dNBS-S and weak ionic associations among –COOH group of alginate molecule, Ca^{2+}, and –COOH group of alginate molecule were employed to build weak ionic cross-linking microstructure. The overlapped Tdmax(ii,iii) values of AdNBS-S (LCHAdNBS-S) could be observed at 390°C. With an increasing introduction of dNBS-S to AdNBS-S composite materials with 5wt% $CaCl_2$, some strong ionic associations were employed to build strong mixed ionic cross-linked microstructure. The remarkable high Tdmax(iii) values of AdNBS-S could be observed at 400°C (HCLAdNBS-S). That is, when the high concentrations of $CaCl_2$(5wt%) and dNBS-S were added, some different associations would be enhanced, such as association between –COOH group of alginate molecule and $-NH_2$ group of dNBS-S, associations among –COOH group of alginate molecule, –COOH group of dNBS-S, and Ca^{2+} ion, associations among –COOH group of alginate molecule, –COOH group of alginate molecule, and Ca^{2+} ion, and associations among –COOH group of dNBS-S, –COOH group of dNBS-S, and Ca^{2+} ion.

With an increasing additions of dNBS-S to composite membranes with 1wt% $CaCl_2$, ordinary ionic association between –COOH group of alginate and $-NH_2$ group of dNBS-S was employed to build weak ionic cross-linked microstructure (**Figure 5**). When a large amount of dNBS-S was introduced into the dNBS-S/alginate composite membrane with 1wt% $CaCl_2$, some ordinary ionic associations such as ionic association between –COOH group of alginate molecule and $-NH_2$ group of dNBS-S, ionic between among –COOH group of alginate molecule, Ca^{2+}, and –COOH group of dNBS-S, and ionic between among –COOH group of dNBS-S, Ca^{2+}, and –COOH group of dNBS-S. The remarkable high Tdmax(iii) values of AdNBS-S could be observed at 400°C.

2.3 Natural soft polymeric composite materials with biomimetic 3D microarchitecture containing gelatin, alginate, and decellularized extracellular matrix via supercritical fluid treatments

For specific clinic applications, the gelatin could also be employed to prepare cross-linked porous natural soft polymeric composite materials with biomimetic 3D micro-architecture containing gelatin, alginate, and decellularized nano-bioscaffolds. A series of composite materials with a biomimetic 3D microarchitecture were prepared based on a fixed weight ratio of gelatin, alginate, and decellularized nano-bioscaffolds. An aqueous gelatin(G) solution and an aqueous alginate(A) solution (weight ratio is 2:1) were homogenized thoroughly. A mixed aqueous solution of gelatin(G) and alginate(A) was obtained as an aqueous GA solution. Briefly, the desired amount of decellularized

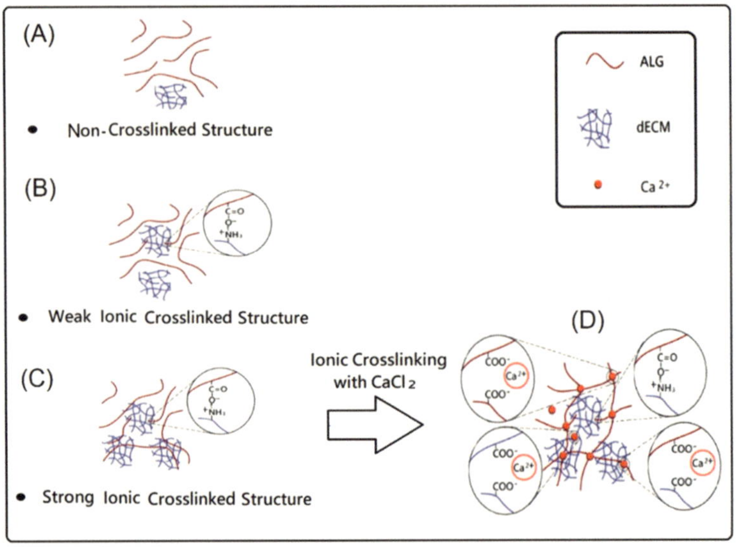

Figure 5.
Proposed model of ionic cross-linked structure from Ca^{2+} ions, alginate molecules, and dNBS-S. (A) non-cross-linked structure, (B) weak ionic cross-linked structure, (C) weak strong ionic cross-linked structure, (D) ionic cross-linked interpenetrating structure.

nano-bioscaffolds powder was first dispersed completely in water. Then, the aqueous GA solution was homogenized thoroughly with the dispersed decellularized nano-bioscaffolds solution. A mixed aqueous solution of gelatin/alginate(GA) and decellularized nano-bioscaffolds was then molded and frozen and then lyophilized. Polymeric porous natural soft polymeric composite materials(GA/dNBS) with biomimetic 3D microarchitecture containing gelatin, alginate, and decellularized nano-bioscaffolds were obtained. The resulting GA/dNBS was further soaked in $CaCl_2$ or GTA aqueous solution with various soaking times for cross-linking reaction with magnet mixer. The cross-linked GA/dNBS solutions were then frozen and dried. A series of designed cross-linked porous composite materials with biomimetic 3D microarchitecture containing gelatin, alginate, and decellularized nano-bioscaffolds could be obtained [32].

2.4 Natural soft polymeric composite materials with biomimetic 3D microarchitecture containing bovine Achilles tendon type I collagen and decellularized extracellular matrix via supercritical fluid treatments

The bovine Achilles tendon type I collagen was dispersed in an acetic acid solution and stirred to get a full dispersion collagen gel. The decellularized extracellular matrix immersed in acetic acid solution at the same concentration with collagen gel resolution was respectively placed on the table concentrator of 25 and 37°C for 24 h, allowing decellularized extracellular matrix to fully swell in acetic acid solution to obtain a decellularized extracellular matrix gel. Afterward, the decellularized extracellular matrix gels were added into the collagen gel following continuous stirring to obtain the composite hydrogel with a final ratio of 9:1 (wt/wt) of collagen and decellularized extracellular matrix. The composite gel was lyophilized. The composite nanobioscaffolds were obtained. After, those scaffolds were fully cross-linked by immersion in a 0.1% (v/v) glutaraldehyde solution (75% ethanol aqueous solution) for 1 h, freeze-dried, and sterilized at a dose of 25 kGy. Finally, the cross-linked collagen/decellularized fibrous extracellular matrix composite nano-bioscaffolds were obtained.

3. The 3D bioprinting applications of natural soft polymeric materials with biomimetic 3D microarchitecture

3.1 Preparation of natural soft polymeric scaffolds with biomimetic 3D microarchitecture by extrusion-based bioprinting procedure with/without cells

Preparation of natural soft polymeric scaffolds with biomimetic 3D microarchitecture could be carried out from dNBS-S and ALG by using an extrusion-based bioprinting procedure with/without cells (**Figure 6**). The uncross-linked natural soft polymeric scaffolds having biomimetic 3D microarchitectures (**Figure 6(A)**) were obtained. Subsequently, the cross-linked interpenetrating composite membranes with/without cells could be obtained after ionic cross-linking procedures (**Figure 6(B)**). On the other hand, the cross-linked interpenetrating composite membranes with/without cells (**Figure 6(D)**) could be directly obtained by using a dual extrusion-based bioprinting procedure (**Figure 6(C)**). Further, introduction of gelatin in designed natural soft polymeric materials with biomimetic 3D microarchitecture could enhance structural and thermal stabilities for tissue engineering and 3D bioprinting applications as a scaffold (**Figure 7**) [32].

3.2 Design of natural soft polymeric materials with biomimetic 3D microarchitecture for dNBS-containing skin equivalents by using a layer-by-layer-dispensing method as active cell-containing wound dressings

Recently, skin grafts could be fabricated using a commercially available bioprinter(CELLINK) [33]. PLG mech was used as a mech to build up a skin construct. For medical application of skin grafts, the designed alginate-based

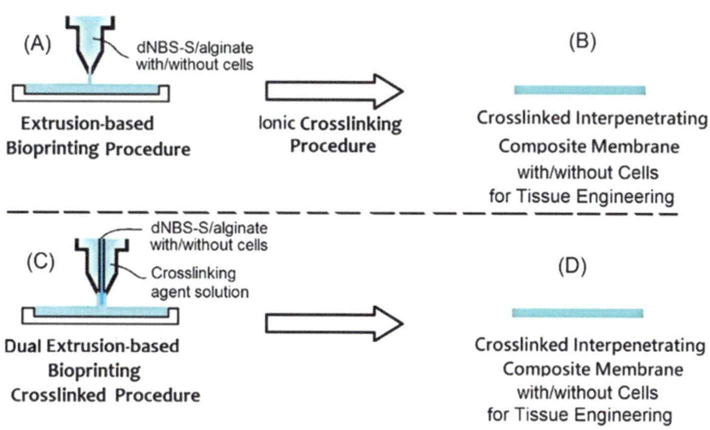

Figure 6.
Proposed model of ionic cross-linked structure from Ca^{2+} ions, alginate molecules, and dNBS-S/alginate. (A) Non-cross-linked structure, (B) weak ionic cross-linked structure, (C) weak strong ionic cross-linked structure, (D) ionic cross-linked interpenetrating structure.

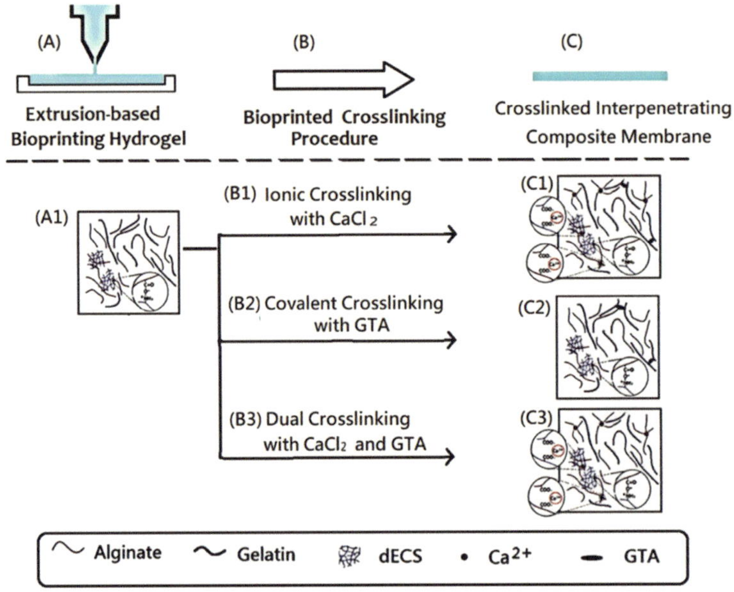

Figure 7.
Proposed model of ionic cross-linked structure from Ca^{2+} ions, alginate molecules, and dNBS-S/alginate. (A) Non-cross-linked structure, (B) weak ionic cross-linked structure, (C) weak strong ionic cross-linked structure, (D) ionic cross-linked interpenetrating structure.

biopolymeric mech containing dECM could be employed to displace PLG mech. The dECM-containing skin equivalents with a dermal layer could be designed using a layer-by-layer-dispensing method. In the concept, a designed dermal construct could be generated by dispensing dermal bioink containing human endothelial cells (ECs), fibroblasts (FBs), and pericytes (PCs) and transfer these cells such as FBs, PCs, and ECs dispensed from a syringe into a mold at an extrusion pressure. A sterile alginate-based biopolymeric mesh was added on top of the first printed dermal layer and incubated at 37°C. The alginate-based biopolymeric mesh was sterilized prior to printing by immersion in a 70% ethanol solution for 30 min and allowed to dry for 1 h at room temperature. Upon gelation, additional dermal bioink containing FBs, PCs, and ECs was printed on top of the mesh, incubated at 37°C until complete gelation, and submerged in media. The resulting construct was further submerged in a larger volume of media to prevent rapid exhaustion of nutrients. Medium was changed daily. After 4 days under medium submersion to allow self-assembly of endothelial networks, the dermal construct was carefully transferred for clinic applications such as an active cell-containing wound dressing (**Figure 8(A)** and **(B)**) or further was bioprinted an additional keratinocytes (KCs) layer to obtain an active KCs-containing wound dressings for wound management (**Figure 8(C)** and **(D)**).

3.3 Design of natural soft polymeric materials with biomimetic 3D microarchitecture for dNBS-containing skin equivalents by using a whole-dispensing method

For specific medical application of skin grafts, the dECS-containing skin equivalents with a dermal layer could be designed using a *whole*-dispensing method. In the concept,

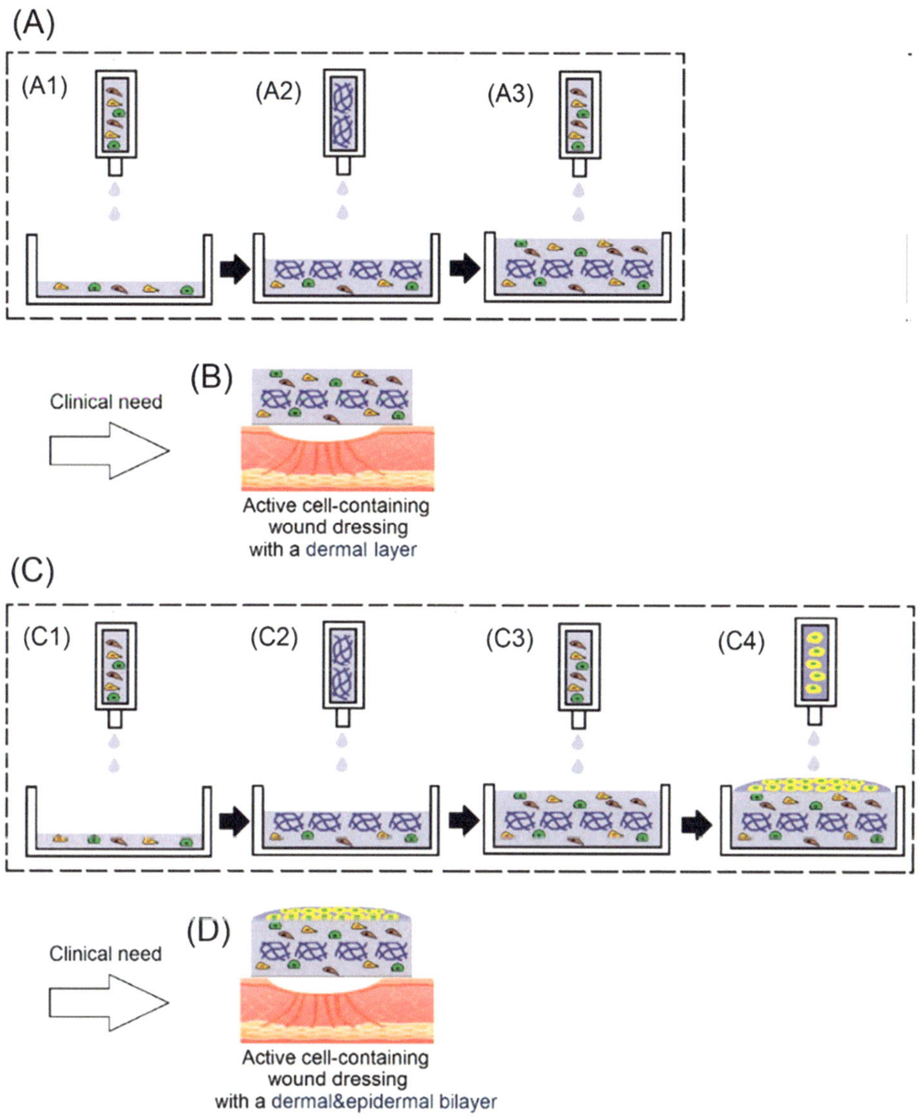

Figure 8.
Proposed model of active cell-containing and active keratinocytes-containing wound dressings for wound management. (A) preparation of an active cell-containing wound dressing, (B) clinical application of an active cell-containing wound dressing, (C) preparation of an active keratinocytes-containing wound dressing, and (D) clinical application of an active keratinocytes-containing wound dressing.

a designed dermal construct could be generated by dispensing dermal bioink containing human endothelial cells (EC), fibroblasts (FBs), pericytes (PCs) , and dECS. The dermal bioink was transferred from a syringe into a mold by using extrusion-based bioprining. A bioprinted dermal layer was obtained and incubated at 37°C (**Figure 9**). The resulting dECS-containing dermal construct was further submerged in a larger volume of media to prevent rapid exhaustion of nutrients. Medium was changed daily. After 4 days under medium submersion to allow self-assembly of endothelial networks, the dECS-containing dermal construct was carefully transferred for clinic applications such as an active dECS/cell-containing wound dressings (**Figure 9(A)**) or further

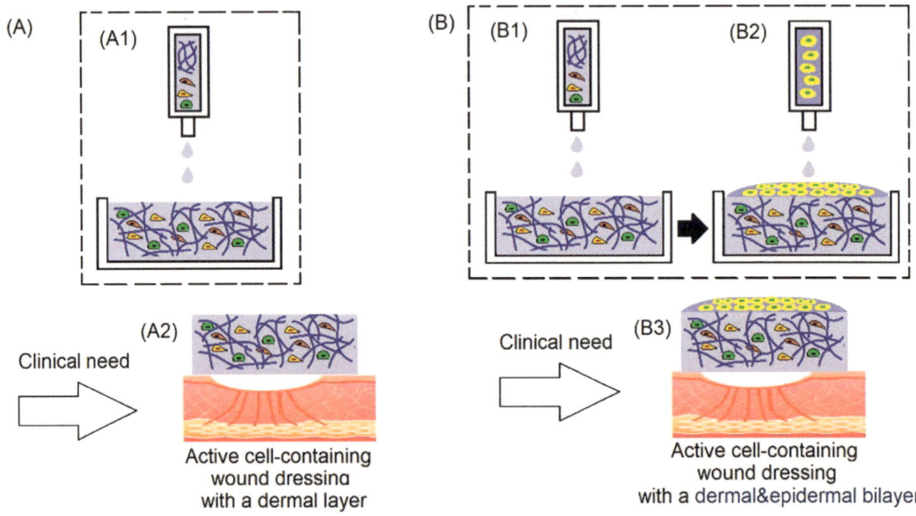

Figure 9.
Proposed model of (A) an active dECS/cell-containing wound dressings and (B) an active dECS/KCs-containing wound dressings.

was bioprinted an additional keratinocytes (KCs) layer to obtain an active dECS/KCs-containing wound dressings for wound management (**Figure 9(B)**).

4. Design of synthetic soft polymeric materials with biomimetic 3D with 3D bioinspired structural foam-wall microarchitectures

For regenerative applications, existing biomedical membranes such as collagen membranes have demonstrated the ability to promote or inhibit cell proliferation. Pure collagens suffer from uncontrollable rapid degradation and weak mechanical strength [34–37]. Generally, synthetic polymers have better mechanical properties and stability than collagen [38]. Polyvinyl alcohol(PVA), which is an FDA-approved material, was widely used in various biomedical applications, including surgical sponges, osteochondral grafts, contact lenses, artificial blood vessels, and implantable medical devices [39, 40] (e.g., bone regeneration [41], wound healing [42], and dental applications [43]) because of the desirable properties such as biocompatibility, non-degradability, low protein absorption, and easily tunable mechanical properties. PVA matrix was characterized by its super water absorbent property, great durability and cleaning ability, and its super soft texture when moist. Using sulfuric acid as catalyst and a suitable pore-forming agent, this porous PVA foam was prepared through PVA acetalization. In usual, some pore-forming agents, such as starch, surfactants, or the reagents [44, 45], which could be employed to produce gas during the cross-linking reaction to prepare PVA foam matrix. Also, water was used as a pore-forming agent to obtain porous structure without any other additional pore-forming agents [46]. Starch was a pore-forming agent that had been commonly used in the preparation of porous PVA matrix [47, 48]. The average pore diameter of PVA foam varied from 30 to 60 μm when wheat-starch was used and varied from 60 to 100 μm when potato starch was used [49, 50]. However, the volumes of PVA matrix gradually shrunk with the increasing acetalization degree and water was continuously excluded during the acetalization process as observation. The PVA matrix with compacted or closed cell porous structure and a poor interconnectivity was obtained. The resulting PVA foam had residual pore-forming agent and undesirable degradable products, which might be harmful for biomedical applications such as tissue engineering or wound management [50]. The PVA/collagen composite materials were successfully developed in the forms of patches, nanofibers, hydrogels, polymer blend, mixed foam, and dual layers [17–55], which were mainly employed for osteochondral defects application [51], tissue-engineered corneas [52], cartilage tissue engineering [53], and wound healing [54, 55]. PVA was employed to modulate the mechanical properties, degradation properties, and cell regulation ability of PVA-based composite materials, which would show the ability to regulate the cell adhesion and proliferation behavior of cell and play a critical role in tissue regeneration [56, 57]. With a strong focus on therapeutic applications, the characteristics of composite materials for medical or regenerative applications examine the inspirations from nature [58, 59].

4.1 Design of synthetic soft polyvinyl alcohol materials with 3D bioinspired structural foam-wall microarchitectures

Biomimetic designs would bring effective materials that are sources of inspiration to biomedical engineers. For tissue engineering and medical applications, a novel bioinspired soft matrix with supporting interconnective foam-wall microarchitectures and/or struts made of cross-linked polyvinyl alcohol with air cavities inspired by avian skeleton and feather rachises was designed. The fully open-cell microstructures with air cavities, structural foam-walls, and structural pneumaticity bioinspired by avian feather rachises and pneumatic bone could be designed. The foam-wall showed a "foam-in-a-foam" microstructure and provided internal reinforcements and pneumaticity [60, 61].

Also, the corresponding foaming process was established. The bioinspired soft matrix with high interconnectivity was fabricated by using a designed air stream pore-forming process [34], which could regulate different pore sizes and microstructure of resulting materials as shown in **Figure 10**. Furthermore, the bioinspired synthetic soft polyvinyl alcohol material was employed to prepare a series of biological modified synthetic soft polyvinyl alcohol materials to provide new functional characteristics for specific clinic and regenerative applications.

The traditional designs of polyvinyl alcohol matrix were prepared by traditional starch pore-foaming process or air-assisted starch pore-foaming process. The medical drainage materials with fully open-cell microstructure could not be obtained. It is difficult to form air cavities to provide interconnectivity and structural support. To build up the stable air cavities, foam walls, and structural pneumaticity bioinspired by avian feather rachises, the introduction of clean atmospheric flow in the foaming process of polyvinyl alcohol matrix would be considered (**Figure 11**). The clean air was incorporated during the pore-foaming process and cross-linking process for formation of air cavities, foam walls, and structural pneumaticity bioinspired by avian feather rachises and pneumatic bone, which could be considered as a biomimetic airstream pore-foaming process. The clean air current was employed to avoid impurity. Complete cross-linking reaction for preparation of polyvinyl alcohol foam materials was also important. During starch-containing pore-foaming process, the pore-foaming agent, starch, could not provide enough driving force to form atmospheric flow, which would promote formation of air cavities with structural support, foam walls, and structural pneumaticity. New biomimetic design of PVA matrix with

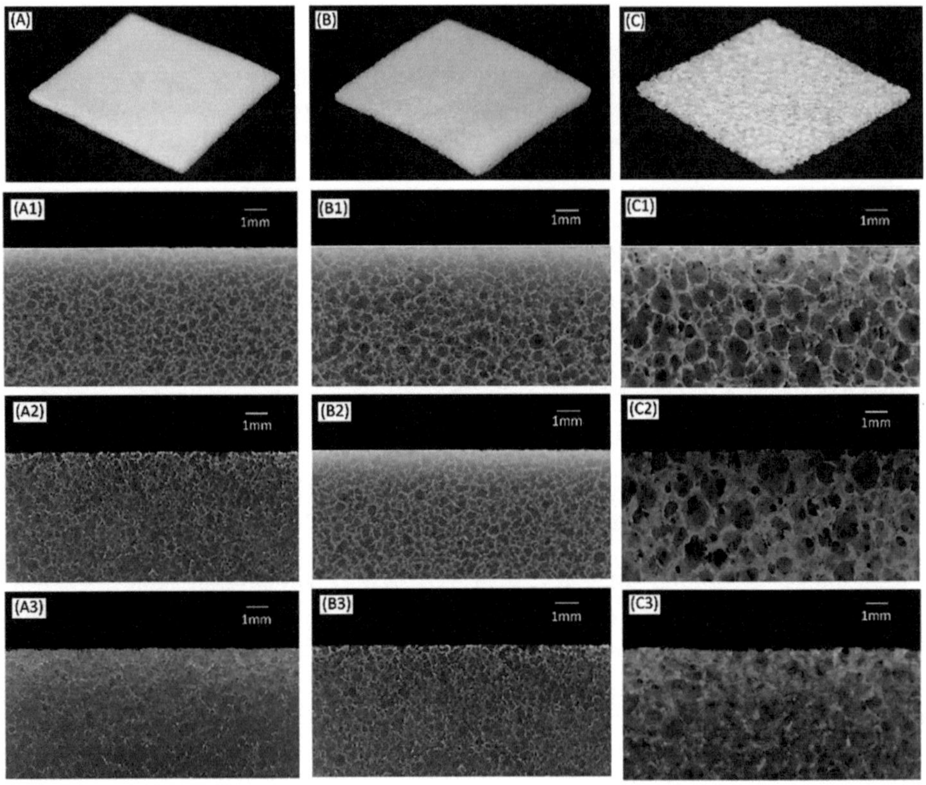

Figure 10.
Macroscopic images of designed synthetic soft polyvinyl alcohol materials with different pore sizes and microstructure. (A) Diameters were ca. 300:(A1) dried type, (A2) subhumid type, (A3) moist type; (B) Diameters were ca. 500:(B1) dried type, (B2) subhumid type,(B3) moist type; and (C) Diameters were ca. 900:(C1) dried type, (C2) subhumid type, (C3) moist type.

air cavities inspired by avian skeleton and feather rachises was prepared by using a designed air flow pore-foaming process without starch pore-foaming agent. A novel b-PVA with fully open cells and channels could be designed and prepared by using a clean airstream pore-foaming process. The reduced pressure could be employed to build up a clean airstream pore-foaming environment. Furthermore, a morphological evaluation of the resulting was carried by using SEM as shown in **Figure 6**. The resulting PVA matrix exhibited spongy structure with fully open-cell interconnecting porous network.

Porous matrices served as a guide for tissue regeneration as a three-dimensional substrate with temporary mechanical support for cell attachment, proliferation, infiltration [62]. The morphology was the key feature that affects both biological and mechanical efficiency of the medical matrices, which would be needed to provide a porous architecture with high interconnectivity to enable cell infiltration, nutrient flow, and integration of the material within the host tissue [63]. Various matrix manufacturing techniques such as emulsion templating [63], gas foaming of heterogeneous blends [64], electrospinning [65], and additive manufacturing [66] had been widely used to introduce porosity into tissue engineering scaffolds. However, PVA matrix always exhibited no interconnectivity or limited interconnectivity

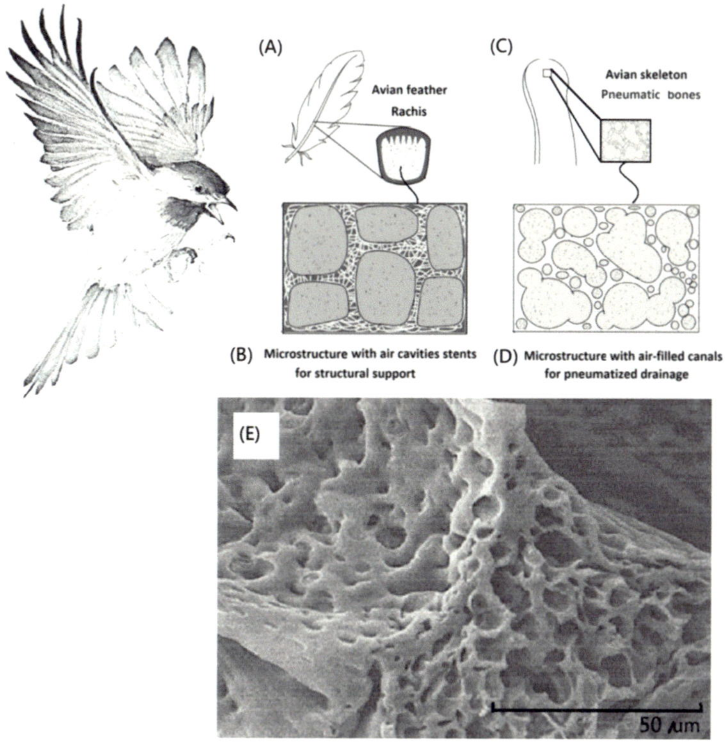

Figure 11.
Microstructures of new biomimetic design of bioinspired PVA matrix were inspired by avian skeleton and feather rachis containing foam-walls. (A) avian feather rachis, (B) microstructure with air cavities stents for structural support, (C) avian skeleton pneumatic bones, (D) microstructure with air-filled canals for pneumatized drainage, and (E) morphological evaluation of new biomimetic design of bioinspired PVA matrix by using scanning electron microscopy (SEM).

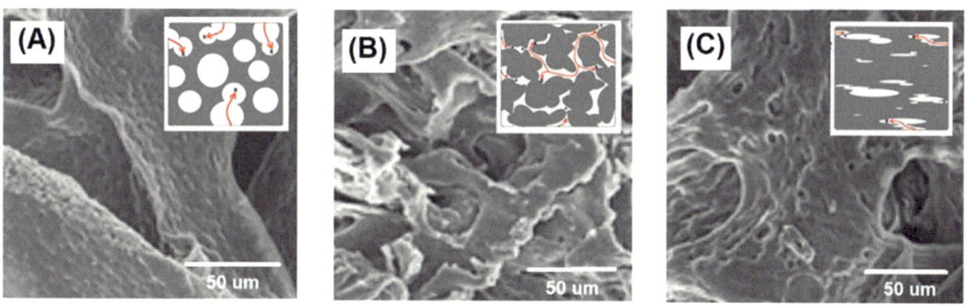

Figure 12.
The microstructural morphological evaluations of commercial soft materials with limited interconnectivity, (A) blind pore, (B) compacted pore, and (C) inaccessible pore.

because of inaccessible pore, blind pore, and compacted pore within the microstructure (**Figure 12**), which might be due to poor foaming ability by using traditional starch pore-foaming process, air-assisted traditional starch pore-foaming process, or incomplete cross-linking reaction of PVA. Biomimetic airstream pore-foaming process was employed as a new matrix fabrication technique to prepare a bioinspired polyvinyl alcohol (b-PVA) with a foam-wall microstructure, which exhibited high porosity and high interconnectivity. The resulting high porosity and interconnectivity would enable cell migration, vascularization, and providing space for newly forming tissues to meet the requirements of specific tissue engineering applications and clinic treatments [66–68].

To understand advanced thermal stability of the synthetic soft polymeric materials with 3D bioinspired structural foam-wall microarchitectures matrix(b-PVA), the commercial medical PVA matrix was also studied by using TGA and DSC and compared with the results of b-PVA. The weight loss curves obtained from thermogravimetric analysis (TGA) could provide some information of thermal degradation such as small molecules, solvent, water, residual reagents, residual foaming agents, starch, weak structural molecules, and prepolymeric molecules, to identify the thermal structural stability. Most of commercial medical PVA matrix exhibited poor thermal stability, which might be due to the compacted and closed-cell microstructure [69]. With increasing temperature, several thermally degraded products and the weight loss of materials were observed. From TGA and DTG results of commercial medical PVA matrix, three stages of weight loss, such as stage I in 50–100°C, stage II in 100–250°C, and stage III in 250–500°C. First, weight loss of commercial medical PVA matrix in stage I at around 100°C was contributed to the loss of water molecules trapped in the hydrophilic PVA. The weight losses of stage I were observed in a range of 3–5 wt%. Second, TGA and DTG spectra of commercial medical PVA matrix in stage II exhibited slight thermal hydrolysis signals increasing with temperature, indicating that the materials might have a poor thermal stable structure and impurity. The weight losses of stage II were determined to be in a range of 2–9 wt%. The relative very low values of initial hydrolysis temperature in stage II were observed at 110–160°C, indicating the thermal degradation of the commercial medical PVA matrix would be happened easily and harm seriously for biomedical applications such as biocompatibility and tissue engineering. Third, TGA and DTG spectra of commercial medical PVA soft matrix in stage III exhibited quite broad peak of DTG curve and the relative low values of initial hydrolysis temperature in stage III were observed in a range of 220–250°C, indicating commercial medical PVA soft matrix might be incomplete cross-linked structures with poor thermal structural stability and incomplete entanglement microstructure with closed cells [34]. Differential scanning calorimeter analysis (DSC) of the commercial medical PVA matrix exhibited some peak below 100°C such as 56, 65, 68, 75, and 95°C, which might be due to the residual compounds such as unreacted PVA or starch pore-forming agents [70]. In general, the conventional method of manufacturing PVA matrix by using the pore-forming agents such as wheat starch (61.8°C [71], 56.2°C [72], and 68.2°C [73]), pea starch (59.8°C [72]), corn starch (71.2 and 80.0°C [71]), potato starch (66.6°C [71], 63.0°C [72], and 72.2°C [73]), cassava starch (65.1°C [73]), tapioca starch (69.9°C [73]), maize starch (76.7°C [73]), and rice starch (68.3°C [71] and 67.8°C/75.3°C [72]). The DSC values of the commercial medical PVA matrix observed at 84–110°C might be contributed to amylopectin part in starch. In the case of the amylopectin part in sago starch, the melting temperature peak was between 50 and 150 °C [74]. Most of the commercial products with compacted and closed cell microstructure were determined and characterized; the results might be contributed to the evident of residual starch after starch pore-foaming process, which could not provide a good support and drainage microstructure. The addition of starch and thermal depredated compounds would be harmful to the clinical applications and risks of pollution in storage system of the resulting soft medical drainage material.

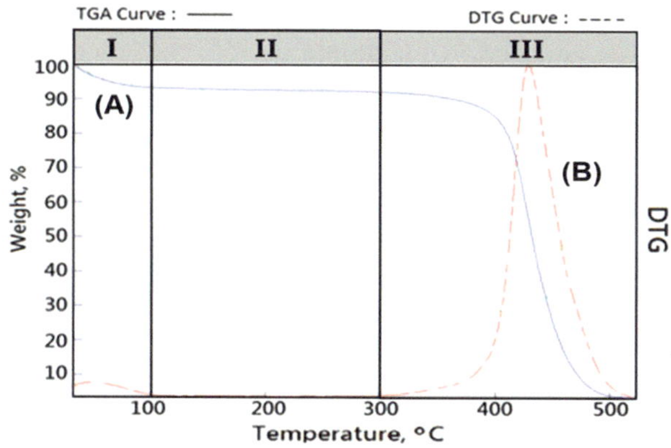

Figure 13.
Thermal characteristics of new bioinspired PVA(b-PVA) matrix, (A) TGA and (B) DTG.

Also, the residual starch would be degraded with an increasing temperature and enhance risks of treatments.

From **Figure 13**, weight loss curves obtained from thermogravimetric analysis (TGA) of biomimetic PVAF could exhibit excellent thermal structural stability and high purity. First, TGA and DTG curves of bioinspired PVA matrix in the region with the temperature lower than 100°C exhibited 6wt% of weight loss, which would be due to only water molecules escaping from the materials and a good water-absorption property. Also, DSC curve of bioinspired PVA matrix was employed to identify thermal stability and exhibited a smooth signal below 100°C, which was contributed to escaping water molecules. Second, TGA and DTG curves of bioinspired PVA matrix in the region with a temperature range between 100 and 300°C exhibited no change of thermal signals, indicating high purity of synthetic soft materials. Third, TGA and DTG curves of bioinspired PVA matrix in the region with the temperature higher than 300°C exhibited a high Tdmax value >425°C and a narrow peak of DTG curve, which indicated high thermal and structural stability and a fully cross-linked structure (**Figure 13**).

A novel bioinspired soft matrix derived from PVA was designed and prepared. Fourier-transform infrared spectra of the designed synthetic soft polymeric materials with 3D bioinspired structural foam-wall microarchitectures (b-PVA) were determined. The main bands of pure PVA at 3500–3400, 2917, 1425, 1324, and 839 cm^{-1} were assigned to the O-H stretching vibration of the hydroxy group, CH_2 asymmetric stretching vibration, C-H bending vibration of CH_2, C-H deformation vibration, and C-C stretching vibration by using FTIR. After the biomimetic airstream pore-foaming process, the molecular structure of PVA matrix was also characterized. The bands at 3500–3400cm^{-1} were weakened and shifted toward higher frequencies due to the consumption of -OH (due to condensation reaction be acetalization tween PVA and formaldehyde) and the corresponding cleavage of the intra- and intermolecular hydrogen bonding. The new bands at 2952, 2913, 2861, and 2675 cm−1 ascribed to symmetric stretching vibrations of the alkyl CH_2 group, the new bands at 1239, 1172, 1129, and 1065 cm^{-1} ascribed to the stretching vibration of C-O in C-O-H groups, and a new peak at 1008 cm^{-1} ascribed to -C-O-C-O-C- stretching vibrations could confirm the formation of a formal structure of b-PVA [75, 76]. Starch of corn, cassava, and potato showed similar absorption bands below 1200 cm^{-1} [77, 78]. A absorbance band around 1150 cm^{-1}, bands at 1080 and 1020 cm^{-1}, and bands around 700–900 cm^{-1} were contributed to vibrations of the glycosidic C–O–C bond [34, 76], the anhydroglucose ring O–C stretch [34, 77], and C-O-C ring vibration [34, 76], respectively [34, 76, 77]. The absorbance band at 924 cm^{-1} was assigned to vibrational modes of a skeletal glycoside bonds of starches [34, 77]. FTIR spectrum of commercial medical PVA matrix showed the similar absorbance band at 924 cm^{-1}. However, the absorbance band at 924 cm^{-1} could not be observed in the FTIR spectra of new bioinspired PVA matrix. The absorbance band at 924 cm^{-1} might be an important evident for starch pore-foaming process and an evaluation for a structural support cell microstructure. The fingerprint region at wavenumbers 1200–600 cm^{-1} could be employed to identify the purity and molecular structure of suitable medical soft matrix. That is, the fingerprint region of new biomimetic design of medical soft matrix could identify the biomimetic super-clean airstream pore-foaming process different from those of traditional starch pore-foaming process. The residual starch pore-foaming agents were degraded easily with an increasing temperature, which might be harmful to the cleanliness of the medical soft matrix and enhance risks of pollution in biomedical application such as tissue engineering or wound management [79, 80].

For the therapeutic application of liver, bioartificial liver support system has been proposed to support the regeneration of the patient's liver [81, 82]. The PVA matrix might be considered as a suitable material. The b-PVA was a nontoxic material with favorable mechanical properties, chemical stability, and good cell adhesion, which could be easily processed and has a special three-dimensional porous structure [83, 84]. Fabrication of porous matrix as potential hepatocyte carriers for bioartificial liver support had been widely explored [85, 86]. Previous studies have tried to immobilize protein with biomaterials by using covalent binding for HepG2 cell culture [85, 86]. The designed b-PVA matrix showed a microstructure containing foam-walls with good structural support and interconnectivity, which might act as a feasible material for artificial liver devices because of biocompatibility, pore-foaming, and mechanical property [85, 86].

The human hepatoblastoma HepG2 cells could be employed for evaluation of liver therapeutic application by using an extrusion-based bioprinting procedure as shown in **Figure 14A**). The ability of cells to adhere and proliferate on the bioinspired polyvinyl alcohol (b-PVA) matrix was an important indicator for evaluating the in vivo application potential. The SEM images were used to evaluate the cell morphology on the bioinspired polyvinyl alcohol matrix after various times of human hepatoblastoma HepG2 cell culture such as 72 h as shown in **Figure 15**. After 72 h of culturing, HepG2 cells would be adhered to the scaffold and grew cover on the interface and inside of the open-cell pores. The foam-walls might provide a rough surface for HepG2 cell proliferation. The interconnected microporous microstructure of the matrix might well supports the penetration of HepG2 cells.

4.2 Design of biological modified b-PVA with 3D bioinspired structural foam-wall microarchitectures

The designed bioinspired PVA(b-PVA) was employed to form a relative biocompatible surface of a collagen-modified b-PVA, which might provide more suitable characteristics for cell adhesion and growth. Some domains in collagen molecules, such as Phe-Hyp-Gly segment, could act as ligands for the integrin family

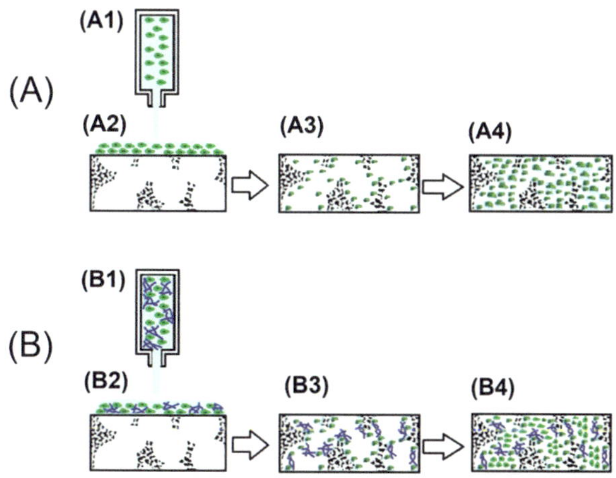

Figure 14.
Bioprinting applications of novel active liver-repairing membranes derived from synthetic soft polymeric materials with biomimetic 3D microarchitecture. (A) Active liver tissue-repairing membranes and (B) active dNBS-containing liver tissue-repairing membranes.

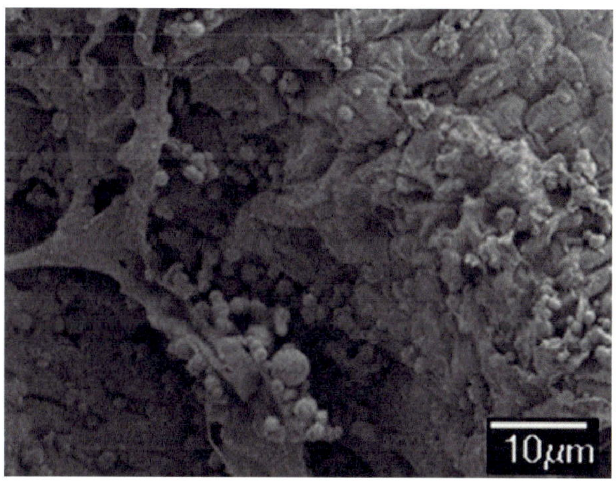

Figure 15.
SEM photographs of HepG2 cells grew on the b-PVA after 72 h (scale bar x10 μm).

receptors of cell surface and potentially activate specific biological signals to promote cell attachment and proliferation. The fully open cell and open foam-wall microstructure of b-PVA was maintained after preparation of collagen-modified b-PVA (**Figure 16**). The b-PVA and bovine Achilles tendon type I collagen were employed to prepare a bioinspired bovine Achilles tendon type I collagen-modified b-PVA composite matrix with open-cell foam-wall microarchitectures. The hydrophilic hydroxyl group within b-PVA molecules might provide a good binding with the coated bovine Achilles tendon type I collagen molecules by using hydrogen bonding and Van der Waals forces [87, 88]. Similar behavior was observed by Zhou et al. [89]. The interaction between b-PVA and bovine Achilles tendon type I collagen molecules within collagen-modified b-PVA was based on non-covalent interactions such as hydrogen

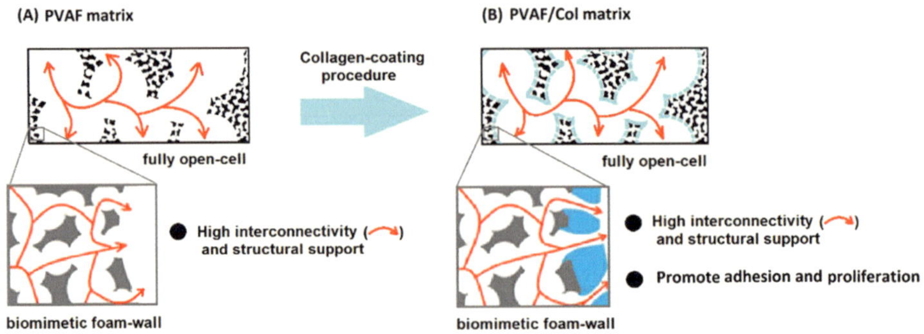

Figure 16.
Interconnected microporous microstructure of medical drainage materials, (A) the original bioinspired PVA matrix and (B) the bioinspired collagen-modified PVA composite matrix.

bonding and Van der Waals forces. The hydrogen bonding would be the main force between PVA and collagen interactions, in which collagen molecule might act as a hydrogen donor and forms hydrogen bonds with the hydroxyl group of b-PVA [90]. The bioinspired soft matrix combined the structural properties of the synthetic PVA matrix with the high cell affinity of the type I collagen, thus enhancing the matrix cytocompatibility. Collagen was the dominant component of the extracellular matrix and had been widely used in tissue engineering due to its low antigenicity, good biocompatibility, biodegradability, and nontoxicity [91]. The FTIR analysis was employed to confirm changes of molecular structure after the surface modification on the PVA matrix with collagen molecules. The absorbances at ~3300, ~2930, ~1636, ~1571, and ~1150 cm^{-1} were observed and characterized for amide A (N-H stretching), amide B(the asymmetrical stretching of CH_2 vibration), amide I(hydrogen bonding between N-H stretching and C=O), amide II(N-H bending and C-N stretching), and amide III(C-N bending, and N-H stretching), respectively. Remarkably, the collagen-modified bioinspired PVA composite matrix showed quite different absorbance from original bioinspired PVA matrix in the region of 1200–1800 cm^{-1}. After the biological surface modification on the b-PVA with collagen molecules, the resulting biological collagen-modified b-PVA still sustained a favorable biomimetic interconnected porous structure.

The morphology of the bioinspired collagen-modified b-PVA with high interconnectivity was investigated and confirmed. After 72 h of culturing, HepG2 cells were remarkably adhered to the bioinspired collagen-modified b-PVA and grew cover on the interface and inside of the open-cell porous microstructure. The SEM images intuitively displayed the growth and distribution of cells on the resulting bioinspired and collagen-modified b-PVA. HepG2 cells adhering to the type I collagen-modified b-PVA showed filopodia, which indicated that human hepatoblastoma HepG2 cells on type I collagen-modified b-PVA in a relative better proliferative condition than pure b-PVA. The stable adhesion and complete spreading could be observed in **Figure 17**. Majhy et al. reported that facile surface modification of substrate could enhance the cell-substrate interaction from a non-adherent state for promoting cell culture [62, 92]. The HepG2 cells attached and grew in cluster on the bioinspired and collagen-modified b-PVA, which was quite different from original b-PVA. The bioinspired and collagen-modified b-PVA was conducive to HepG2 proliferation, migration, expression, and functionality maintenance.

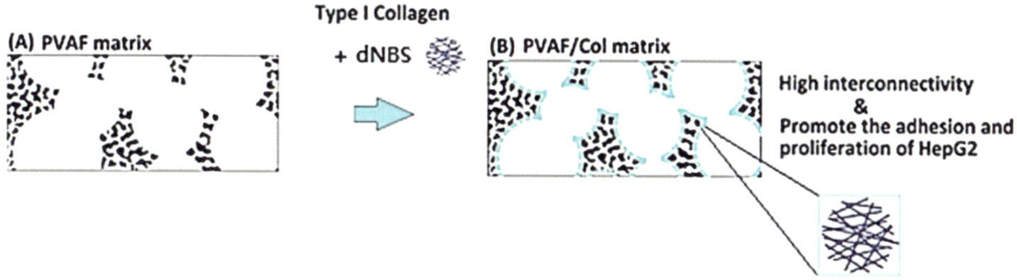

Figure 17.
Design of (A) bioinspired PVA composite matrix and the corresponding (B) dNBS-modified biological bioinspired PVA composite matrix.

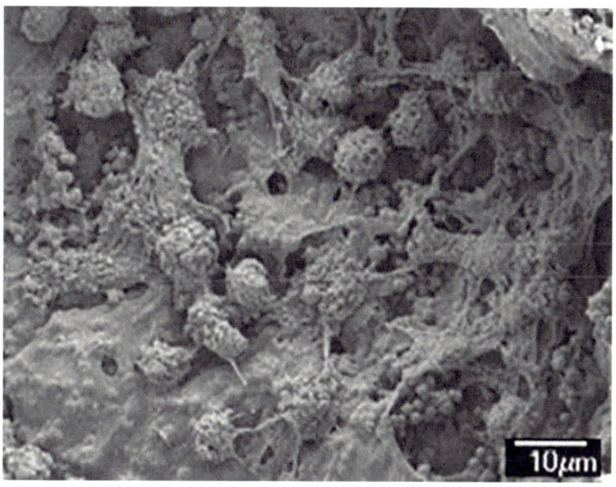

Figure 18.
SEM photographs of HepG2 cells grew on the collagen modified b-PVA (scale bar x10 μm).

HepG2 cells proliferated actively and formed cell clusters more efficiently in the collagen-modified b-PVA than original b-PVA (**Figure 18**). The collagen of collagen-modified b-PVA appeared to promote the growth and differentiation of the HepG2 cells. As shown in **Figure 18**, after 72 h, cells attached and grew well on the collagen-modified PVA composite matrix and a large number of pseudopods could be seen. The HepG2 cells would be evenly distributed on the surface of the pore of the b-PVA. Similarly, Moscato et al. reported that poly(vinyl alcohol)/gelatin hydrogels were good for growth of HepG2 cells [63, 93]. In the future, the bioinspired and collagen-modified b-PVA might serve as a promising liver cell culture carrier to be used in the biological artificial liver reactor. Both the type I collagen-modified b-PVA and the corresponding b-PVA could be considered as good materials for different tissue engineering applications. Furthermore, the type I collagen could be combined with dNBS to prepare new designed collagen/dNBS-modified b-PVA, which might provide additional 3D microenvironments within porous structure for cell adhesion and growth as shown in **Figure 17**. Also, the human hepatoblastoma HepG2 cells could be employed for evaluation of liver therapeutic application such as active liver-tissue-repairing membranes by using an extrusion-based bioprinting procedure as shown in **Figure 14(B)**.

4.3 Design of synthetic chitosan oligosaccharides-modified PVA composite matrix(COS/b-PVA) with 3D bioinspired structural foam-wall microarchitectures

Chitosan oligosaccharide (COS) is a biomaterial obtained by chemical or enzymatic degradation of chitosan derived from shrimp and crab shells [94]. The resulting chitosan oligosaccharide(COS) is composed of 2–10 glucosamines linked by β-1, 4 glycosidic bonds and the molecular weight is up to 3900 Da. The b-PVA was soaked in chitosan oligosaccharide (COS) solution and then frozen-dried. Synthetic chitosan oligosaccharides-modified b-PVA composite matrix(COS/b-PVA) with 3D bioinspired structural foam-wall microarchitectures was obtained. The open-cell microstructures were retained in chitosan oligosaccharides-modified polyvinyl alcohol(COS/b-PVA) (**Figure 19**). The connecting holes and the corresponding open channels were fully covered and filled with chitosan oligosaccharides, which provide a controlled release system of active chitosan oligosaccharide molecules for wound managements. The water permeability of the resulting chitosan oligosaccharide-modified COS/b-PVA dressings was determined by ASTM D4491 standard test methods. The good water permeability was observed in a relative higher level than 80%, which would be contributed to specific clinic applications [94–96].

4.4 Design of soft polymeric materials with 3D bioinspired structural foam-wall microarchitectures using a mixed foaming procedure combined synthetic polyvinyl alcohol with natural chitosan

A new form of polymer blend, macroporous chitosan/poly(vinyl alcohol) (PVA) foams made by a starch expansion process, exhibits the functionalities of chitosan while avoiding its poor mechanical properties and chemical instabilities. The appropriate conditions for foaming are discussed using both insoluble and water-soluble chitosan. Both insoluble/PVA and water-soluble chitosan/PVA foams demonstrated interconnected and open-cell structures with large pore size from tens to hundreds of micrometers and high porosities from 73.6 to 84.3%. The chitosan/PVA with high interconnected and open-cell structures might have potential in the tissue

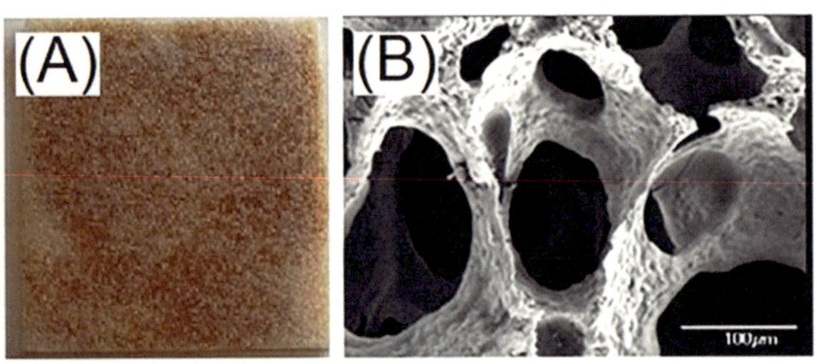

Figure 19.
(A) Photo of the designed permeable chitosan oligosaccharide modified polyvinyl alcohol complex foam (COS/b-PVA) and (B) morphology of designed chitosan oligosaccharide modified polyvinyl alcohol complex foam (COS/b-PVA) dressing with fully open-cell and open-channel microstructures.

Figure 20.
Active tissue-repairing sponges made from chitosan/poly(vinyl alcohol) (Supported by Cenefom and PARSD).

engineering and bioprinting applications [97]. The clean airstream pore-foaming process for b-PVA could be employed to build up a blend-foaming process for preparation of chitosan/poly(vinyl alcohol) (b-PVA). The chitosan/poly(vinyl alcohol) (PVA) was obtained and applied for clinic treatments such as active tissue-repairing sponges (**Figure 20**).

5. Clinic application of active tissue-repairing membranes by using 3D bioprinting and cells culture procedure

New biomimetic biomedical inventive design-thinking methods could be established for design of new clinic application such as active tissue-repairing membranes by using 3D bioprinting and cells culture procedure [98].

5.1 Clinic application of synthetic soft active tissue-repairing membranes by using 3D bioprinting and cells culture procedure

The collagen-coating PVA could be employed as a 3D-bioprinting supporting scaffold for 3D bioprinting applications of a tissue-repairing membrane such as active skin-tissue-repairing membranes and liver-repairing membranes.

The collagen-coating b-PVA could be cultured HepG2 cell to design a liver-repairing membrane as shown in **Figure 15**. Also, the bioprinting of a human skin substitute containing xeno-free cultured human EC, FBs, and PCs in a xeno-free bioink (**Figure 21(A)(A1)**) containing human collagen type I and fibronectin layered in a biocompatible collagen-coating b-PVA supporting scaffold (**Figure 21(A)(A2)**). After cells of EC, FBs, and PCs being cultured and transferred into the biocompatible collagen-coating b-PVA supporting scaffold (**Figure 21(A)(A3)**), a designed active skin-tissue-repairing membranes containing human EC, FBs, and PCs could be obtained for clinical treatments (**Figure 21(A)(A4)**). Furthermore, dNBS could be introduced into the bioink (**Figure 21(B)(B1)**) containing EC, FBs, PCs, and dNBS, human collagen type I, and fibronectin and layered in a biocompatible

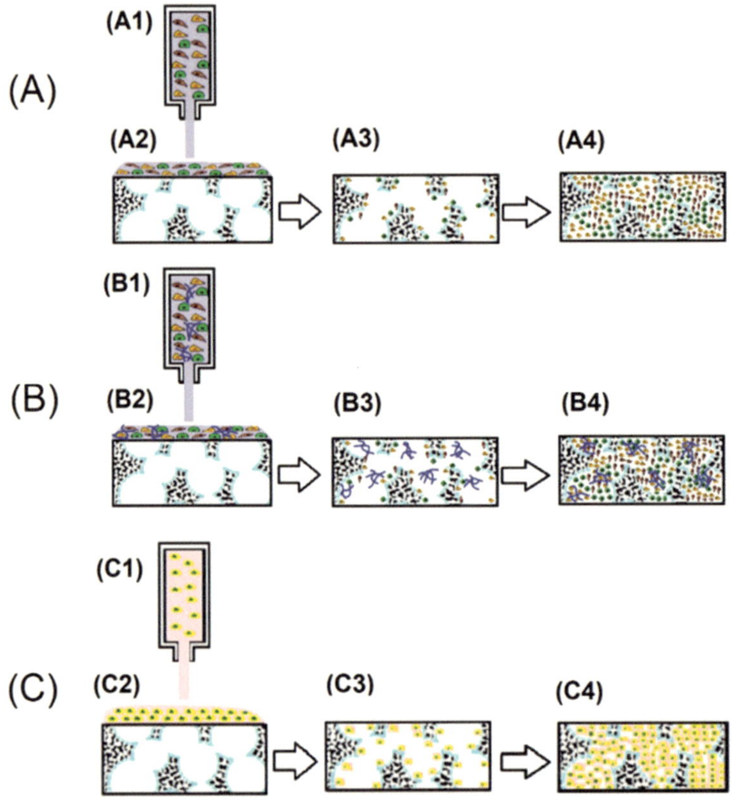

Figure 21.
Bioprinting applications of novel active skin-repairing membranes derived from natural and synthetic soft polymeric materials with biomimetic 3D microarchitecture. (A) active skin dermal tissue-repairing membranes, (B) active dNBS-containing skin dermal tissue-repairing membranes, and (C) active skin epidermal tissue-repairing membranes.

collagen-coating b-PVA supporting scaffold (**Figure 20(B)(B2)**). The collagen-coating b-PVA could be cultured EC, FBs, and PCs with dNBS (**Figure 20(B)(B3)**) to obtain a novel active dNBS-containing skin-tissue-repairing membranes containing human EC, FBs, PCs, and dNBS (**Figure 21(B)(B4)**). Independently, human KCs were introduced into a xeno-free bioink (**Figure 21(C)(C1)**). Furthermore, the collagen-coated PVA could be bioprinted with the resulting KCs bioinks (**Figure 20(C)(C2)**) and cultured with xeno-free human KCs (**Figure 21(C)**(C3)) to form active epidermal tissue-repairing membranes for clinic application in wound management (**Figure 21(C)(C4)**).

For the clinic applications, the novel active skin dermal tissue-repairing membranes, dNBS-containing active skin dermal tissue-repairing membranes, and active skin epidermal tissue-repairing membranes could be applied directly on the target wounds for different repairing steps such as dermal repairing (**Figures 22(A)** and **23(A)**) and subsequently epidermal repairing (**Figures 22(C)** and **23(C)**). Also, the additional PBS or saline could be employed to transfer the cells within the wound dressings onto the surface of wound depending on the clinic needs for different repairing steps such as dermal repairing (**Figures 22(B)** and **23(B)** and subsequently epidermal repairing (**Figures 22(D)** and **23D)**). The dNBS-containing active skin repairing wound dressings could provide a natural biomimetic 3D microarchitecture

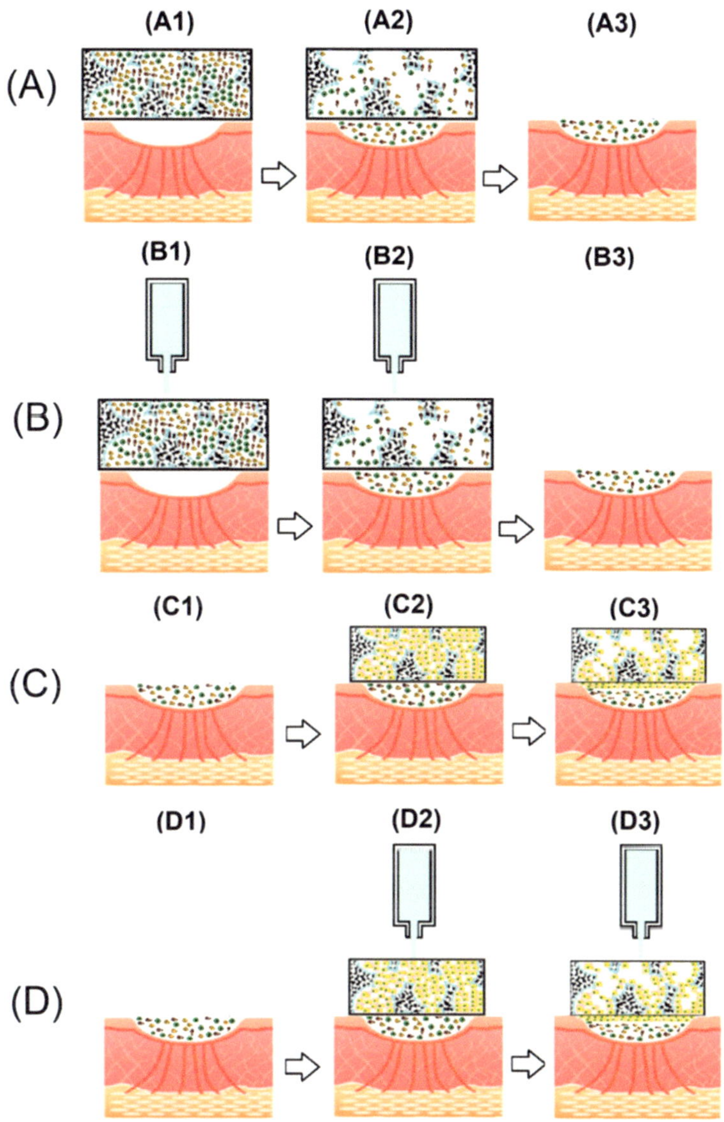

Figure 22.
Novel active skin-repairing membranes derived from synthetics oft polymeric materials with biomimetic 3D microarchitecture for wound managements. (A) Active skin dermal tissue-repairing membranes without transferring assistances, (B) active skin dermal tissue-repairing membranes with transferring assistances, (C) active skin epidermal tissue-repairing membranes without transferring assistances, and (D) active skin epidermal tissue-repairing membranes with transferring assistances.

for promotion of cell growth. The collagen-coating PVA matrix was considered as a good cell carrier for cell culturing and transferring.

6. Conclusions

Design and characterization of natural and synthetic soft polymeric materials with biomimetic 3D microarchitecture were considered. The natural soft polymeric

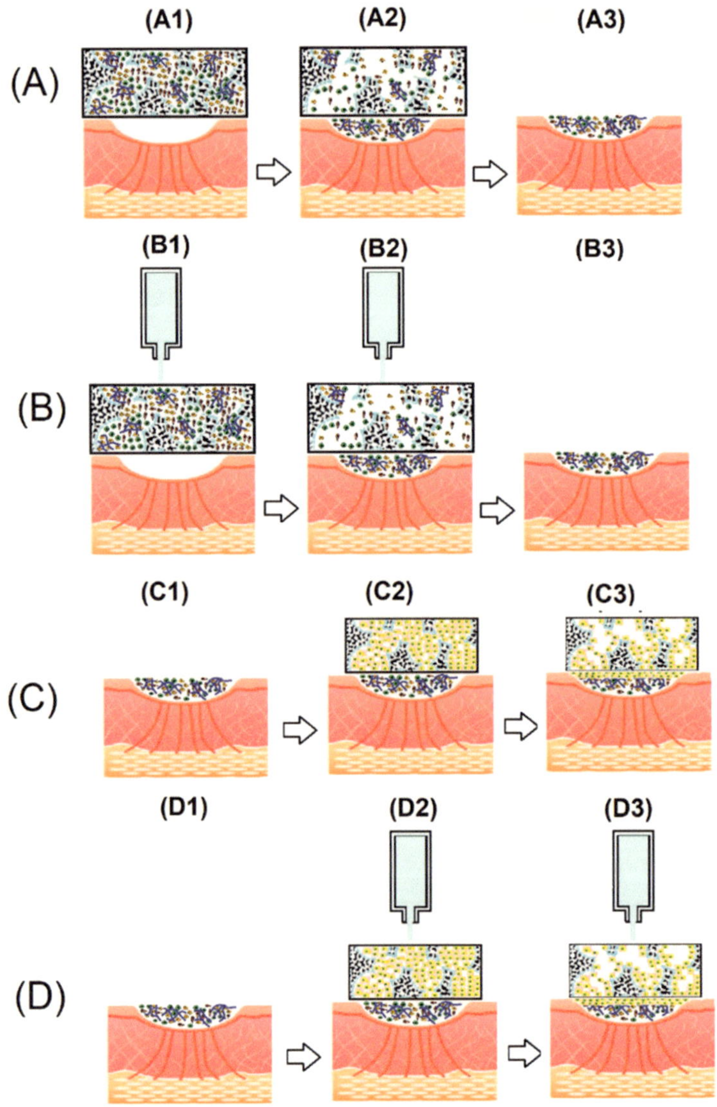

Figure 23.
Novel active dNBS-containing skin-repairing membranes derived from synthetic soft polymeric materials with biomimetic 3D microarchitecture for wound managements. (A) active dNBS-containing skin dermal tissue-repairing membranes without transferring assistances, (B) active dNBS-containing skin dermal tissue-repairing membranes with transferring assistances, (C) active dNBS-containing skin epidermal tissue-repairing membranes without transferring assistances, and (D) active dNBS-containing skin epidermal tissue-repairing membranes with transferring assistances.

materials would focus on new design bioinspired membranes containing supercritical fluids-decellularized dermal scaffolds for 3D bioprinting potential applications. Synthetic soft polymeric materials would focus on b-PVA with structural foam-wall microarchitectures. The b-PVA was designed, prepared, and showed relative higher interconnectivity than commercial medical soft matrix, which could be considered as a good potential scaffold for tissue engineering. Furthermore, the b-PVA was employed to prepare a highly biocompatible collagen-modified b-PVA composite matrix after surface modification with type I collagen. In the future, the collagen-modified b-PVA composite matrix might serve as a promising liver cell culture carrier to be employed in the biological artificial liver reactor.

Acknowledgements

Author would like to acknowledge the Taiwan PARSD Pharmaceutical Technology Consultants Ltd. Company for financial and technical support. Additionally, the author would like to thank Gui-Feng Zhang and Yuguang Du for their technical assistance from the State Key Laboratory of Biochemical Engineering, Institute of Process Engineering, and Chinese Academy of Sciences, Beijing.

Conflicts of interest

The authors declare no conflict of interest.

Author details

Ching-Cheng Huang[1,2*] and Masashi Shiotsuki[3]

1 Department of Biomedical Engineering, Ming-Chuan University, Taiwan

2 PARSD Biomedical Material Research Center, Taiwan

3 Faculty of Science and Engineering, Tokyo City University, Japan

*Address all correspondence to: jcchuang.mcu@gmail.com

References

[1] Murphy CA, Costa JB, Silva-Correia J, Oliveira JM, Reis RL, Collins MN. Biopolymers and polymers in the search of alternative treatments for meniscal regeneration: State of the art and future trends. Applied Materials Today. 2018;**12**:51-71

[2] Kwon H, Brown WE, Lee CA. Surgical and tissue engineering strategies for articular cartilage and meniscus repair. Nature Reviews Rheumatology. 2019;**15**(9):550-570

[3] Bilgen B, Jayasuriya CT, Owens BD. Current concepts in Meniscus Tissue engineering and repair. Advanced Healthcare Materials. 2018;**7**(11):1701407

[4] Rongen JJ, van Tienen TG, van Bochove B, Grijpma DW, Buma P. Biomaterials in search of a meniscus substitute. Biomaterials. 2014;**35**(11):3527-3540

[5] Deo KA, Singh KA, Peak CW, Alge DL, Gaharwar AK. Bioprinting 101: Design, fabrication, and evaluation of Cell-Laden 3D Bioprinted Scaffolds. Tissue Engineering. Part A. 2020;**26**(5-6):318-338

[6] Murphy SV, Atala A. 3D bioprinting of tissues and organs. Nature Biotechnology. 2014;**32**(8):773-785

[7] Filardo G, Petretta M, Cavallo C. Patient-specific meniscus prototype based on 3D bioprinting of human cell-laden scaffold. Bone Joint Research. 2019;**8**(2):101-106

[8] Cengiz IF, Pitikakis M, Cesario L. Building the basis for patient-specific meniscal scaffolds: From human knee MRI to fabrication of 3D printed scaffolds. Bioprinting. 2016;**1-2**:1-10

[9] Mishra A, Srivastava V. Biomaterials and 3D printing techniques used in the medical field. Journal of Medical Engineering & Technology. 2021;**5**(4):290-302

[10] Zhang J, Wehrle E, Rubert M, Müller R. 3D bioprinting of human tissues: Biofabrication, Bioinks, and Bioreactors. International Journal of Molecular Sciences. 2021;**22**(8):3971

[11] Ravnic DJ, Leberfinger AN, Koduru SV, Hospodiuk M, Moncal KK, Datta P, et al. Transplantation of bioprinted tissues and organs: Technical and clinical challenges and future perspectives. Annals of Surgery. 2017;**266**(1):48-58

[12] Li J, Chen M, Fan X, Zhou H. Recent advances in bioprinting techniques: Approaches, applications and future prospects. Journal of Translational Medicine. 2016;**14**:271

[13] Chen H, Zhang Y, Zhang L. Applications of bioinspired approaches and challenges in medical devices. Bio-des. Manufacturing. 2021;**4**:146-148

[14] Li Z, Huang S, Liu Y, Yao B, Hu T, Shi H, et al. Tuning alginate-gelatin bioink properties by varying solvent and their impact on stem cell behavior. Scientific Reports. 2018;**8**:8020

[15] Brovold M, Almeida JI, Pla-Palacín I, Sainz-Arnal P, Sánchez-Romero N, Rivas JJ, et al. Naturally-derived biomaterials for tissue engineering applications. Advances in Experimental Medicine and Biology. 2018;**1077**:421-449

[16] Carpio BM, Dabaghi M, Ungureanu J, Kolb MR, Hirota JA, Moran-Mirabal JM. 3D bioprinting strategies, challenges, and opportunities to model the lung tissue

microenvironment and its function. Frontiers in Bioengineering and Biotechnology. 2021;**9**:773511

[17] Huang CC, Chen YJ, Liu HW. Characterization of composite nano-bioscaffolds based on collagen and supercritical fluids-assisted decellularized fibrous extracellular matrix. Polymers (Basel). 2021;**13**(24):4326

[18] Tappa K, Jammalamadaka U. Novel biomaterials used in medical 3D printing techniques. Journal of Functional Biomaterials. 2018;**9**(1):17

[19] Rhee S, Puetzer JL, Mason BN, Reinhart-King CA, Bonassar LJ. 3D bioprinting of spatially heterogeneous collagen constructs for cartilage tissue engineering. ACS Biomaterials Science & Engineering. 2016;**2**:1800-1805

[20] Laronda MM, Rutz AL, Xiao S, Whelan KA, Duncan FE, Roth EW, et al. A bioprosthetic ovary created using 3D printed microporous scaffolds restores ovarian function in sterilized mice. Nature Communications. 2017;**8**:15261

[21] Huang CC. Decellularized liver-regenerative 3D printing biomaterials for cell-based liver therapies via a designed procedure combining supercritical fluids with papain-containing reagents. Bio-medical Materials and Engineering. 2022;**33**(2):139-146

[22] Markstedt K, Mantas A, Tournier I, Martínez-Ávila H, Hägg D, Gatenholm P. 3D bioprinting human chondrocytes with nanocellulose–alginate bioink for cartilage tissue engineering applications. Biomacromolecules. 2015;**16**:1489-1496

[23] Li S, Tian X, Fan J, Tong H, Ao Q, Wang X. Chitosans for tissue repair and organ three-dimensional (3D) bioprinting. Micromachines (Basel). 2019;**10**(11):765

[24] Wang X, Wang Q, Xu C. Nanocellulose-based inks for 3D bioprinting: Key aspects in research development and challenging perspectives in applications: A mini review. Bioengineering (Basel). 2020;**7**(2):40

[25] Gong C, Kong Z, Wang X. The effect of agarose on 3D bioprinting. Polymers (Basel). 2021;**13**(22):4028

[26] Li JX, Zhao SX, Zhang YQ. Silk protein composite bioinks and their 3D scaffolds and in vitro characterization. International Journal of Molecular Sciences. 2022;**23**(2):910

[27] Nguyen D, Hägg DA, Forsman A, Ekholm J, Nimkingratan P, Brantsing C, et al. Cartilage tissue engineering by the 3D bioprinting of iPS cells in a nanocellulose/alginate bioink. Scientific Reports. 2017;**7**:658

[28] Petta D, D'Amora U, Ambrosio L, Grijpma DW, Eglin D, D'Este M. Hyaluronic acid as a bioink for extrusion-based 3D printing. Biofabrication. 2020;**12**(3):032001

[29] Lam T, Dehne T, Krüger JP, Hondke S, Endres M, Thomas A, et al. Photopolymerizable gelatin and hyaluronic acid for stereolithographic 3D bioprinting of tissue-engineered cartilage. Journal of Biomedical Materials Research. Part B, Applied Biomaterials. 2019;**107**(8):2649-2657

[30] Huang CC. New designed decellularized scaffolds for scaffold-based gene therapy from elastic cartilages via supercritical carbon dioxide fluid and alkaline/protease treatments. Current Gene Therapy. 2021

[31] Perez-Puyana V, Romero A, Guerrero A. Influence of collagen concentration and glutaraldehyde on

collagen-based scaffold properties. Journal of Biomedial Materials Research Part A. 2016;**104**:1462-1468

[32] Huang CC. Tuning gelatin–alginate bioink properties by introducing new decellularized elastic cartilage scaffolds for bioinspired composite membranes in orthopedics. Polymer Bulletin. 2022

[33] Baltazar T, Merola J, Catarino C, Xie CB, Kirkiles-Smith NC, Lee V, et al. Three dimensional bioprinting of a vascularized and perfusable skin graft using human keratinocytes, fibroblasts, pericytes, and endothelial cells. Tissue Engineering. Part A. 2020;**26**(5-6):227-238

[34] Huang CC. Design and characterization of a bioinspired polyvinyl alcohol matrix with structural foam-wall microarchitectures for potential tissue engineering applications. Polymers. 2022;**14**:1585

[35] Kasaj A, Reichert C, Götz H, Röhrig B, Smeets R, Willershausen B. In vitro evaluation of various bioabsorbable and nonresorbable barrier membranes for guided tissue regeneration. Head & Face Medicine. 2008;**4**:1-8

[36] Subhan F, Hussain Z, Tauseef I, Shehzad A, Wahid F. A review on recent advances and applications of fish collagen. Critical Reviews in Food Science and Nutrition. 2020;**61**:1027-1037

[37] Dattola E, Parrotta EI, Scalise S, Perozziello G, Limongi T, Candeloro P, et al. Development of 3D PVA scaffolds for cardiac tissue engineering and cell screening applications. RSC Advances. 2019;**9**:4246-4257

[38] Saavedra YG, Mateescu MA, Averill-Bates DA, Denizeau F. Polyvinylalcohol three-dimensional matrices for improved long-term dynamic culture of hepatocytes. Journal of Biomedical Materials Research. Part A. 2003;**66**:562-570

[39] Baker MI, Walsh SP, Schwartz Z, Boyan BD. A review of polyvinyl alcohol and its uses in cartilage and orthopedic applications. Journal of Biomedical Materials and Research B. 2012;**100**:1451-1457

[40] Hou Y, Xie W, Achazi K, Cuellar-Camacho JL, Melzig MF, Chen W, et al. Injectable degradable PVA microgels prepared by microfluidic technology for controlled osteogenic differentiation of mesenchymal stem cells. Acta Biomaterialia. 2018;**77**:28-37

[41] Chen CY, Kuo SM, Tarng YW, Lin KC. Immediate application of negative pressure wound therapy following lower extremity fap reconstruction in sixteen patients. Scientific Reports. 2021;**11**:21158

[42] Hamidabadi HG, Rezvani Z, Bojnordi MN, Shirinzadeh H, Seifalian AM, et al. Chitosan-intercalated montmorillonite/poly(vinyl alcohol) nanofibers as a platform to guide neuronlike differentiation of human dental pulp stem cells. ACS Applied Materials Interfaces. 2017;**9**:11392-11404

[43] Peppas NA, Stauffer SRJ. Reinforced uncrosslinked poly (vinyl alcohol) gels produced by cyclic freezing-thawing processes: A short review. Journal of Controlled Release. 1991;**16**:305-310

[44] Trimnell D, Shasha BS, Otey FHJ. The effect of α-amylases upon the release of trifluralin encapsulated in starch. Journal of Controlled Release. 1985;**1**:183-190

[45] Chang YI, Cheng WY, Jang L. A novel method of making PVF Porous foam without using the pore forming

agent. Journal of Applied Polymer Science. 2014;**132**:41270

[46] BeMiller JN, Whistler RL. Starch: Chemistry and Technology. 3rd ed. New York, NY, USA: Academic Press; 2009

[47] Wilson JD, Bechtel DB, Todd TC, Seib PA. Cereal Chemistry. 2006;**83**:259

[48] 48.Lan W, Xu M, Qin M, Cheng Y, Zhao Y, Huang D, et al. Physicochemical properties and biocompatibility of the bi-layer polyvinyl alcohol-based hydrogel for osteochondral tissue engineering. Materials and Design. 2021;**204**:109652

[49] Wu Z, Kong B. Engineering of corneal tissue through an aligned PVA/collagen composite nanofibrous electrospun scaffold. Nanomaterials. 2018;**8**:124-140

[50] 50.Barbon S, Contran M, Stocco E, Todros S, Macchi V, Caro RD, et al. Enhanced biomechanical properties of polyvinyl alcohol-based hybrid scaffolds for cartilage tissue engineering. PRO. 2021;**9**:730

[51] Iqbal B, Muhammad N, Rahim A, Iqbal F, Sharif F, Safi SZ, et al. Development of collagen/PVA composites patches for osteochondral defects using a green processing of ionic liquid. International Journal of Polymeric Materials. 2019;**68**:590-596

[52] Choi SM, Singh D, Kumar A, Oh TH, Cho YW, Han SS. Porous three-dimensional PVA/gelatin sponge for skin tissue engineering. International Journal of Polymeric Materials and Polymeric Biomaterials. 2013;**62**:384-389

[53] Kamoun EA, Kenawy ER, Chen SX. A review on polymeric hydrogel membranes for wound dressing applications: PVA-based hydrogel dressings. Journal of Advanced Research. 2017;**8**:217-233

[54] Sionkowska A. Current research on the blends of natural and synthetic polymers as new biomaterials: Review. Progress in Polymer Science. 2011;**36**:1254-1276

[55] Min Q, Tian D, Zhang Y, Wang C, Wan Y, Wu J. Strong and elastic chitosan/silk fibroin hydrogels incorporated with growth-factor-loaded microspheres for cartilage tissue engineering. Biomimetics. 2022;**7**:41

[56] Reddy R, Reddy N. Biomimetic approaches for tissue engineering. Journal of Biomaterials Science. Polymer Edition. 2018;**29**:1667-1685

[57] Ngadimin KD, Stokes A, Gentile P, Ferreira AM. Biomimetic hydrogels designed for cartilage tissue engineering. Biomaterials Science. 2021;**9**:4246-4259

[58] Niu W, Xiao Q, Wang X, Zhu J, Li J, Liang X, et al. A biomimetic drug delivery system by integrating grapefruit extracellular vesicles and doxorubicin-loaded heparin-based nanoparticles for glioma therapy. Nano Letters. 2021;**21**:1484-1492

[59] Ziai Y, Petronella F, Rinoldi C. Chameleon-inspired multifunctional plasmonic nanoplatforms for biosensing applications. NPG Asia Materials. 2022;**14**:18

[60] Lingham-Soliar T. Feather structure, biomechanics and biomimetics: The incredible lightness of being. Journal für Ornithologie. 2014;**155**:323-336

[61] Bonser RHC. The mechanical performance of medullary foam from feathers. Journal of Materials Science Letters. 2001;**20**:941-942

[62] Aldemir DB, Claeyssens F. Basic principles of emulsion templating and its use as an emerging manufacturing method of tissue engineering scaffolds. Frontiers in Bioengineering and Biotechnology. 2020;**8**:875

[63] Bak TY, Kook MS, Jung SC, Kim BH. Biological effect of gas plasma treatment on CO_2 gas foaming/salt leaching fabricated porous polycaprolactone scaffolds in bone tissue engineering. Journal of Nanomaterials. 2014;**3**:1-6

[64] Dikici S, Claeyssens F, MacNeil S. Bioengineering vascular networks to study angiogenesis and vascularization of physiologically relevant tissue models in Vitro. ACS Biomaterials Science & Engineering. 2020;**6**:3513-3528

[65] Dikici A, Dikici BS, Karaman O, Oflaz H. The effect of zinc oxide doping on mechanical and biological properties of 3D printed calcium sulfate based scaffolds. Biocybernetics Biomedical Engineering. 2017;**37**:733-741

[66] Richez A, Deleuze H, Vedrenne P, Collier R. Preparation of ultra-low-density microcellular materials. Journal of Applied Polymer Science. 2005;**96**:2053-2063

[67] Loh QL, Choong C. Three-dimensional scaffolds for tissue engineering applications: Role of porosity and pore size. Tissue Engineering. Part B, Reviews. 2013;**19**:485-502

[68] Despa F, Orgill DP, Neuwalder J, Lee RC. The relative thermal stability of tissue macromolecules and cellular structure in burn injury. Burns. 2005;**31**:568-577

[69] Rynkowska E, Fatyeyeva K, Marais S, Kujawa J, Kujawski W. Chemically and thermally crosslinked PVA-based membranes: Effect on swelling and transport behavior. Polymers. 1799;**2019**:11

[70] Obanni M, Bemiller JN. Properties of some starch blends. Cereal Chemistry. 1997;**74**:431-436

[71] Jacobs H. Influence of annealing on the pasting properties of starches from varying botanical sources. Cereal Chemistry. 1995;**72**:480-487

[72] Ratnayake WS, Otani C, Jackson DS. DSC enthalpic transitions during starch gelatinization in excess water, dilute sodium chloride, and dilute sucrose solutions. Journal of the Science of Food and Agriculture. 2009;**89**:2156-2164

[73] Beninca C, Colman TAD, Lacerda LG, Filho MASC, Demiate IM, Bannach G, et al. Thermal, rheological, and structural behaviors of natural and modified cassava starch granules, with sodium hypochlorite solutions. Journal of Analytical Calorimetry. 2013;**111**:2217-2222

[74] Hamdan S, Hashim DMA, Ahmad M, Embong S. Compatibility studies of polypropylene (PP)-sago starch (SS) blends using DMTA. Journal of Polymer Research. 2000;7:237-244

[75] Tao TX, Wu ZC, Wang XQ, Li MS, Zhang JH. Synthesis and spectra of compleses involving polyvinyl alcohol fiber ligands. Acta Polymerica Sinica. 2006;**3**:387-390

[76] Kizil R, Irudayaraj J, Seetharaman K. Characterization of irradiated starches by using FT-Raman and FTIR spectroscopy. Journal of Agricultural and Food Chemistry. 2002;**50**:3912-3918

[77] Dankar I, Haddarah A, Omar FEL, Sepulcre F. Influence of storage under unfavourable conditions on the caking properties and fungal contamination

of potato starch and wheat flour. Food Chemistry. 2018;**260**:7-12

[78] Abdullah AHD, Chalimah S, Primadona I, Hanantyo MHG. Physical and chemical properties of corn, cassava, and potato starchs. Earth and Environmental Science. 2018;**160**:012003

[79] Li J, Zeng H, Zeng Z, Zeng Y, Xie T. Promising graphene-based nanomaterials and their biomedical applications and potential risks: A comprehensive review. ACS Biomaterials Science & Engineering. 2021;7:5363-5396

[80] Sahana; T.G.; Rekha, P.D. Biopolymers: Applications in wound healing and skin tissue engineering. Molecular Biology Reports , 2018, 45, 2857-2867.

[81] Lee JS, Cho SW. Liver tissue engineering: Recent advances in the development of a bio-artificial liver. Biotechnol Bioproc E. 2012;**17**:427-438

[82] Zhang Q, Wang Q, Lv S, Lu J, Jiang S, Regenstein JM, et al. Comparison of collagen and gelatin extracted from the skins of Nile tilapia (Oreochromis niloticus) and channel catfish (Ictalurus punctatus). Food Bioscience. 2016;**13**:41-48

[83] Karimi A, Navidbakhsh M, Beigzadeh B. A visco-hyperelastic constitutive approach for modeling polyvinyl alcohol sponge. Tissue & Cell. 2014;**46**:97-102

[84] Stampella A, Papi A, Rizzitelli G. Synthesis and characterization of a novel poly(vinyl alcohol) 3D platform for the evaluation of hepatocytes response to drug administration. Journal of Materials Chemistry B. 2013;**2**:1-37

[85] Jain E, Damania A, Shakya AK. Fabrication of macroporous cryogels as potential hepatocyte carriers for bioartificial liver support. Colloids and Surfaces. B, Biointerfaces. 2015;**136**:761-771

[86] Jiang X, Christopherson GT, Mao HQ. The effect of nanofibre surface amine density and conjugate structure on the adhesion and proliferation of human haematopoietic progenitor cells. Interface Focus. 2011;**1**:725-733

[87] Xu S, Gu M, Wu K, Li G. Unraveling the interaction mechanism between collagen and alcohols with different chain lengths and hydroxyl positions. Colloids and Surfaces. B, Biointerfaces. 2021;**199**:111559

[88] Lai G, Du Z, Li G. The rheological behavior of collagen dispersion/poly(vinyl alcohol) blends. Korea-Austin Rheological Journal. 2007;**19**:81-88

[89] Zhou T, Zheng K, Sui B, Boccaccini AR, Sun J. In vitro evaluation of poly (vinyl alcohol)/collagen blended hydrogels for regulating human periodontal ligament fibroblasts and gingival fibroblasts. International Journal of Biological Macromolecules. 2020;**163**:1938-1946

[90] Lan W, Xu M, Zhang X, Zhao L, Huang D, Wei X, et al. Biomimetic polyvinyl alcohol/type II collagen hydrogels for cartilage tissue engineering. Journal of Biomaterials Science. Polymers. 2020;**31**:1-22

[91] Asran AS, Henning S, Michler GH. Polyvinyl alcohol–collagen–hydroxyapatite biocomposite nanofibrous scaffold: Mimicking the key features of natural bone at the nanoscale level. Polymer. 2010;**51**:868-876

[92] Majhy B, Priyadarshini P, Sen AK. Effect of surface energy and roughness on cell adhesion and growth–facile

surface modification for enhanced cell culture. RSC Advances. 2021;**11**:15467-15476

[93] Moscato S, Ronca F, Campani D. Poly(vinyl alcohol)/gelatin hydrogels cultured with HepG2 Cells as a 3D model of hepatocellular carcinoma: A morphological study. Journal of Functional Biomaterials. 2015;**6**:16-32

[94] Lei X, Zhang G, Yang MJ, Du Y, Huang CC. Morphology of novel chitosan oligosaccharides modified cross-linked polyvinyl alcohol complex foam (COS/cPVACF) dressings with fully open-cell and open-channel microstructures via active molecules cleaning process for wound managements. Biomedical Journal of Scientific & Technical Research. 2019;**17**(3):12819-12823

[95] Huang CC, Wu XW, Hsieh YL, Lin MH, Li J, Du Y, et al. Preparation and preclinical evaluations of deigned polymeric chitosanoligosaccharide (COS)/chitosan composite microcapsules with uniform sizes as a novel control released carrier via premixed membrane emulsification. Basic & Clinical Pharmacology & Toxicology. 2020;**126**(1):66

[96] Huang CC, Lin MH, Hsieh YL, Wu XW, Li J, Du Y, et al. Production and preclinical evaluation of designed chitosanoligosaccharide surface-modified microcapsules with uniform size in aqueous dispersions as a novel control released carrier via premixed membrane emulsification. Basic & Clinical Pharmacology & Toxicology. 2020;**126**(1):67

[97] Wang X, Chung YS, Lyoo WS, Min BG. Preparation and properties of chitosan/poly(vinyl alcohol) blend foams for copper adsorption. Polymer International. 2006;**55**(11):1230-1235

[98] Huang CC. Establishment of BM-TRIZ biomedical inventive principles and design-thinking methods for innovative design of medical devices based on a new polymeric biomaterial containing polyvinyl alcohol foam via an air-foaming procedure. Annals of Advanced Biomedical Sciences (AABSc). 2021;**4**(1):1-12

Chapter 4

Novel Composites for Bone Tissue Engineering

Pugalanthipandian Sankaralingam, Poornimadevi Sakthivel and Vijayakumar Chinnaswamy Thangavel

Abstract

Novel metal oxide-doped fluorophosphates nano-glass powders were synthesized by melt quenching method, and their non-toxicity is proved by MTT. Their efficacy in bone formation is confirmed by osteocalcin and ALP secretion. Composites were made using PLA, PDLLA, PPF, or 1,2-diol with fluorophosphates nano-glass powders (AgFp/MgFp/ZnFp). Their non-toxicity was assessed by cell adhesion and MTT. The ability of the composite for bioconversion was assessed by RT-PCR estimation for osteocalcin, Collagen II, RUNX2, Chondroitin sulfate, and ALP secretion accessed by ELISA method. The animal study in rabbit showed good callus formation by bioconduction and bioinduction. The bioconversion of the composite itself was proved by modified Tetrachrome staining. From the 12 different composites with different composition, the composite PPF+PDLLA+PPF+ZnFp showed the best results. These obtained results of the composites made from common biological molecules are better than the standards and so they do biomimic as bone substitutes. The composites can be made as strips or granules or cylinders and will be a boon to the operating surgeon. The composite meets nearly all the requirements for bone tissue engineering and nullifies the defect in the existing ceramic composites.

Keywords: fluorophosphate, nano bioactive glass, bioconversion, bone substitute, synthetic bone graft

1. Introduction

Scientists have tried natural polymer, synthetic polymer, various ceramics, and composites, etc., as bone graft substitutes. An ideal bone substitute should be non-toxic, osteoconductive, osteoinductive, bioconvertable, physical properties should be near to that of normal bone, should be cost-effective, easily sterilizable, and be feasible for bulk production. The commonly available bone graft substitutes do not meet all the above requirements [1, 2].

The materials that were put to use initially were ceramics (hydroxyapatite and tricalcium phosphate) as bone graft substitutes. Hydroxyapatite was only osteoconductive and rarely was converted to bone even after years [3, 4]. It was not useful in replacing weight-bearing function. Tricalcium phosphate had minimal osteoinductive capacity along with osteoconduction but had no bioconversion capability [5, 6].

To have the advantage of bioconversion, certain specific bone hormones such as "Bone Morphogenic Protein" (BMP) came into use. "Demineralized Bone Matrix" (DBM) was also marketed as bone graft substitute, which was discovered by Urist [7]. The essential problem in their use is the phenomenal cost involved, and they had good osteoinduction but were not good osteoconductors. Hench [8] came out with the 45S5 glass, which was a breakthrough as it was made from cheap chemicals, was osteoconductive as well as osteoinductive, was able to merge with the natural bone, and is commercially available. The drawback with 45S5 glass is their very slow resorption, the longer time taken for bioconversion, and their inability to be used as a weight-bearing implant. The time taken for bioconversion was a drawback in reducing the duration of morbidity of the patients. To circumvent these problems, silica-free phosphate bioglasses and metal-oxide-doped bioglasses entered into these fields.

Over decades, orthopedic surgeons know chronic ingestion of fluoride in abundance leads to fluorosis, an ectopic bone-forming condition [9]. This was used to our advantage, and the ideal concentration of fluoride in the ceramics network was standardized. The evaporation of fluorine from the fluoride compound was circumvented by the new methodology of preparing fluoride bioglasses by melt quenching.

Standardization of the ideal mole percentage of fluoride resulted in the invention of fluorophosphates glasses, which are much more bio active than other types of phosphate and silica glasses and had a higher rate of bioconversion and faster resorption [10, 11]. Doping them with metal oxides improved their physical properties and brought the elastic moduli close to that of the human bone. Scaffolding the fluorophosphate glasses was essential to bring the molecule for clinical use that will bridge the gap between the need of the surgeon and the capability of the scientist.

This novel synthetic composite is made from bioinert polymers comprising poly lactic acid, poly D, L -Lactic acid, and bioactive polymers consisting of polypropylene fumarate, diester of fumaric acid, and 1,2 propylene diol and a bioactive inorganic component consisting of a metal-doped fluorophosphates nano glass powder. It can be fabricated as granules, scaffolds such as strip and cylinder. The plurality of the shapes that can be made gives additional advantage to the surgeon.

The standard biomaterial should mimic the biological process in the microenvironment simply denoted as "Biomimetics." The selected three different metals have beneficial properties as follows: (a) Silver: The best antimicrobial agent is especially in wound healing and skin care. (b) Magnesium: It plays a vital role in bone structure development and also acts as dietary supplement in medicine. (c) Zinc: It involves in DNA synthesis, gene expression, and wound healing. Also, the degradation products of the biocompatible polymers (PLA, PDLLA, PPF, and 1,2-diol) all enter the Kreb's cycle and excreted. So, the metal oxides and the polymers induce bone formation by stimulating the genetic pathway of bone formation and so are biomimetic.

2. Materials

Poly lactic acid (PLA) and poly DL-lactic acid (PDLLA) were procured from BioDegmer® Japan. Polypropylene fumarate (PPF) was prepared by transesterification method [12]. The di-1,2-propanediol ester of fumaric acid (1,2-diol) was prepared by azeotropic distillation, and its detailed explanation is available in previous publication [13].

The fluorophosphate glasses were prepared by melt quenching method and converted to nano particles by mechanical milling. The measured quantities of the required chemicals (Na_2CO_3, $CaCO_3$, CaF_2, P_2O_5, and $ZnO/Ag_2O/MgO$) were taken in a ball mill and homogenized. The mixture was heated in a 10% alumina crucible for 1 h up to 120°C and cooled to room temperature. The material was again ball milled for 1 hr. The components were taken in a platinum crucible and kept in a furnace preheated to 1100°C for 90 min. Then the crucible with the material was quenched by plunging in liquid nitrogen. The formed glass was broken to pieces and milled for 48 h to obtain metal-oxide-doped nano powder of the specific fluorophosphates glass [14].

2.1 Gel foam casting under rapid heating

The required amounts of the PLA [15], PDLLA, PPF/1,2-diol, and AgFP/ZnFP/MgFP were taken and dispersed in dichloromethane using a magnetic stirrer (300 rpm). Once the mixture was homogenized, the material was slowly poured over a hot glass plate (70°C). The rapid evaporation of dichloromethane leads to the generation of random pores in the composite scaffold formed. After complete evaporation of the solvent, highly interconnected porous scaffold with homogeneous distribution of the components was obtained. The scaffold thus made can be ground to granules, cut into strip, or rolled into cylinder [16].

To know the significance of PDLLA, scaffolds were fabricated with or without PDLLA [17]. The fabricated composites are:1. PLA+1,2-diol+AgFp, 2. PLA+PDLLA+1,2-diol+AgFp, 3. PLA+PPF+AgFp, 4. PLA+PDLLA+PPF+AgFp, 5. PLA+1,2-diol+ZnFp, 6. PLA+PDLLA+1,2-diol+ZnFp, 7. PLA+PPF+ZnFp, 8. PLA+PDLLA+PPF+ZnFp, 9. PLA+1,2-diol+MgFp, 10.PLA+PDLLA+1,2-diol+MgFp, 11. PLA+PPF+MgFp, 12. PLA+PDLLA+PPF+MgFp.

3. Methodology

3.1 In vitro evaluation of the scaffolds

3.1.1 Metal-doped fluorophosphates nano bioglasses.

The biological activity (cell viability, attachment, and proliferation) of the metal-doped fluorophophate nano bioglass was ascertained by cell adhesion and MTT assay; Potential osteogenic differentiation was ascertained by (Alkaline Phosphatase Activity) and non-collagenous protein (intra and extracellular osteocalcin) of the nano glass, in relation with MG63 cell lines.

3.1.2 Scaffolds (PLA+PDLLA+PPF/1,2-diol+AgFp/ZnFp/MgFp)

The significance of the pores in the scaffold was assessed by calcein AM study and MTT evaluation. By following standard Kokubo protocol, simulated body fluid (SBF) was prepared. All the fabricated scaffolds were cut into $2X2cm^2$ size. The scaffolds were placed in 20 mL SBF filled glass container, for a period of 21 days at 5%

CO_2 incubator (Heraus – Germany). The pH variation was noted everyday using pH meter E1 model. After 21 days, the scaffolds were carefully removed; dried in laminar air flow for 48 h. The variation in the pH over 21 days was noted. The deposition of hydroxyapatite and fluorapatite was investigated by FTIR and SEM_EDAX. The toxicity of the different composites with different composition of the components was ascertained by the MTT in relation to the SaOS2 and human endothelial cell lines. Their efficiency in enhancing secretion of alkaline phosphatase and chondroitin sulfate, the ground substance of the bone was measured by ELISA method. The ability of the composites in the secretion of osteocalcin, Collagen II, RUNX2 was assessed by RT-PCR method. The porosity and micro-architecture in the multilayered scaffold were assessed by Micro-CT evaluation with GE SRμCT analyzer.

The primers used for PCR were as follows:

- Collagen type II
- Forward primer: CATGAGGGCGCGGTAGAGA.
- Reverse Primer: ATCCCCTCTGGGTCCTTGTT.
- Product length: 296.
- Osteocalcin.
- Forward primer: TCACACTCCTCGCCCTATTG.
- Reverse Primer: CTCTTCACTACCTCGCTGCC.
- Product length: 132.
- RUNX2Sequence (5′- > 3′) Template strand Length Start Stop Tm GC% Self complementarity Self 3′ complementarity.
- Forward primer CCACCGAGACCAACAGAGTC Plus.
- Reverse primer GTCACTGTGCTGAAGAGGCT.
- Product length: 119.

The analysis of the results was performed using ABI PRISM® 7000 Sequence Detection System software that enables more sensitive and accurate estimation of the relative gene expression.

3.2 In vivo studies

3.2.1 For granules

The *in vivo* studies were conducted with the Ethics committee approval (Ethical committee approval no. ABS/IAEC/18-10-2019/003-). A single species of *Orictologuscuniculus* was purchased from King Institute, Chennai, India, and domesticated over a period of 2 weeks. The day-night rhythm was maintained and was fed on good nourishing food. The adaptation was confirmed by the gain in weight of 150 g in 2-week time. (1800–1950 g).The composite (PLA+PDLLA+PPF+AgFp) granules were prepared by grounding the scaffold and sterilized by ethylene oxide gas.

The animal was given a premedication of pedichloryl syrup (2.5 mL) 30 minutes before surgery. Intramuscular ketamine anesthesia was given in the dose of 45 mg per kilogram body weight and waited for 10 minutes to get the full dissociated anesthetic effect. The anesthetic effect was maintained by oxygen and sevoprim inhalation through mask (**Figure 1**).

The left thigh was repeatedly painted with 10% povidone iodine and ethylene alcohol. Xylocaine 2% with adrenaline was injected in the line of incision as an additional analgesia and also a hemostatic agent. The skin incision made on the anterolateral aspect was rolled down to expose the posterior boarder of the quadriceps muscle. Using sharp dissection, the muscle was slit open and enlarged by thin bone spikes to expose the anterolateral aspect of the thigh bone. Using an electric dental burr of 1 mm, a trough was made for a length of 2 cm.This exposed the medullary cavity. It was packed with the sterile composite powder. Liberal saline wash was given to wash off the spilled over composite materials. The spikes once removed the muscle fell back into position completely covering the bony trough. The two 3–0 vicryl stitches were used to close the muscle. The skin incision, which was far away from the bone work, was closed with 3–0 ethilon. A single dose of ceftrioxazone 250 mg was given intramuscularly.

3.2.2 For strip

The *in vivo* studies were conducted with the Ethics committee approval (Ethical committee approval no. ABS/IAEC/18-10-2019/003-). Three male rabbits were

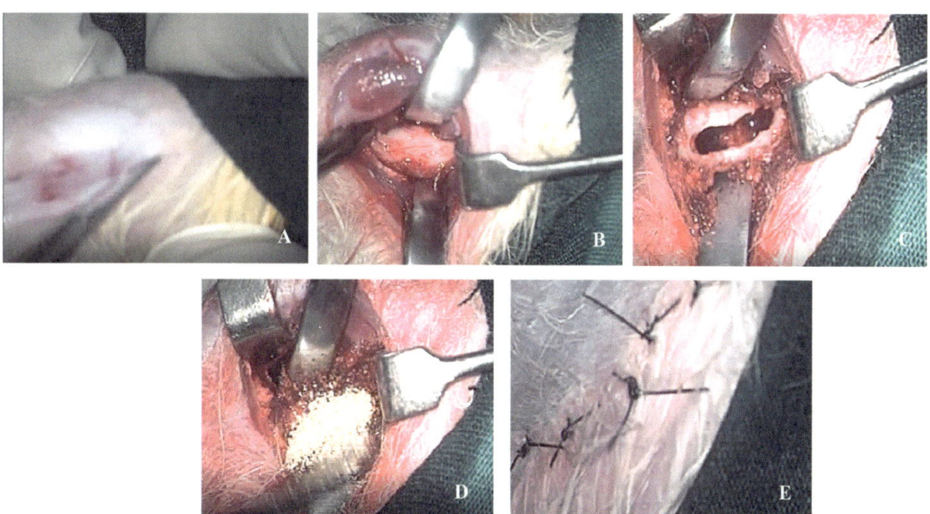

Figure 1.
In vivo study in rabbit (granules-PLA+PDLLA+PPF+AgFp) [A-thigh preparation, B - femur exposed, C - gutter in the femur, D - gutter filled with granules E - Wound closed by sutures.

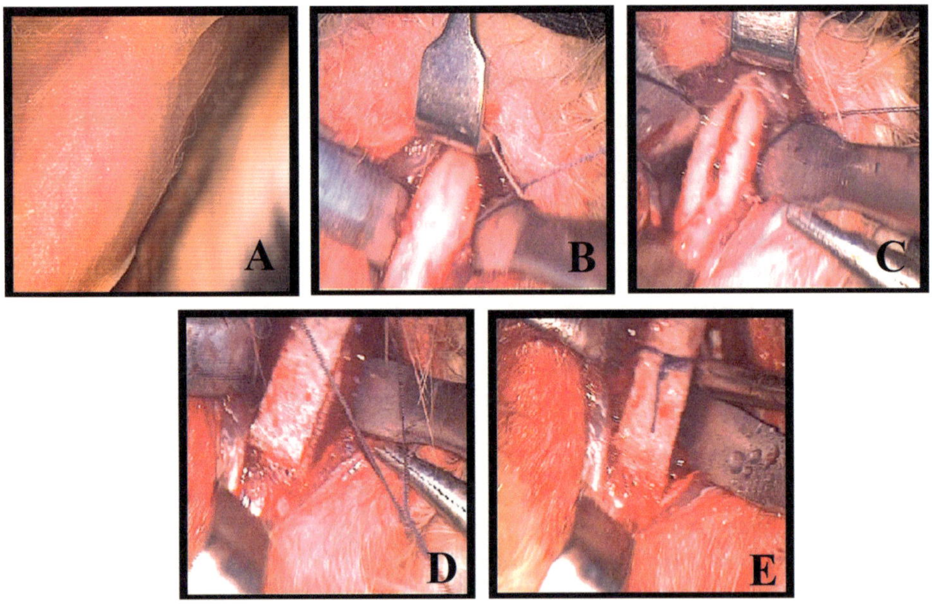

Figure 2.
In-vivo study in rabbits (strip -PLA+PDLLA+PPF+AgFp). [A - Thigh preparation, B - Femur exposed, C - Cut in the femur, D - Two strips one above another over the cut, E - Maintain position by vicryl knot].

procured and domesticated in the same way as explained before. The AgFp/ZnFp/MgFp composites were made with PLA+PDLLA+PPF by gel foam casting under rapid heating. They were of 1 mm thickness and cut into size of 2X20mm. The cut specimens were sterilized by ethylene oxide gas sterilization.

The animals were anesthetized, limb prepared, and femur exposed as described in the previous section. Narrow cuts were made with no.701 conical dental burr at an angle of 45° to the femur to make it extremely thin cut. The 3–0 vicryl was threaded around the femur and both the ends were kept free. Two layers of the 2X20mm sterilized composite strips were kept over the cut made allowing the marrow blood to soak the specimen. The vicryl was tied around the specimen so that the specimen does not slip or move away and the wound was closed in layers. The procedure was done for all the three specimens, one on each animal (**Figure 2**).

The animals were cared for the postoperative period with nourishing food. The day 1 X-ray did not show the specimen in either view as the specimens were translucent to the X-ray. The X-ray evaluation was done under sedation on the first, ninth, and 16th day. Clinical union occurred as early as the 15th day. The CT evaluation was done on the 19th day. The animals were euthanized as per the protocol and the limb harvested, denuded of skin and muscles, and bone preserved in 10% formalin. The X-rays of the specimens taken and then sent for histopathological evaluation in both EH stain and modified tetrachrome stain. The procedure adapted is shown in the serial photographs in **Figure 2**, where two layers of 1 mm thick strips have been placed over a very narrow corticotomy wound in the shaft of femur and have been retained in position by a single 3–0 vicryl encircling knot.

4. Results and discussions

4.1 Metal-doped fluorophosphates nanobioglasses

Alkaline phosphatase (ALP) is an osteogenic differentiation marker at all the stages from the differentiation of the mesenchymal cells to the mineralization front. Hence its enhanced secretion is considered as a vital factor to choose the ingredient for the composite for Bone Tissue Engineering (BTE). The obtained results (**Table 1**) indicated that AgFp and MgFp showed consistently raised levels at all concentration from 0.1 to 100 μg/mL, whereas ZnFp showed increased secretion only at lower concentrations of 0.01 and 1 μg/mL [18, 19].

Bone is a composite of the ground substance reinforced by multiple collagens and mineralized by hydroxyapatite [20]. Though various collagens are present in various parts of the body, osteocalcin is found exclusively in bone. It is also an excellent gene marker of bone induction. The ability of the ionic dissolution products of various FP glasses in various concentrations was evaluated for their efficiency to promote osteocalcin secretion. While the extracellular expression of osteocalcin [21] showed increase than the control only with ZnFp and MgFp, (**Table 2**) intracellular osteocalcin was raised in most of the glasses, but significant raise was present in ZnFp, MgFp, and AgFp glasses and was more when the concentration of the products of dissolution was 10 μg/mL.

The calcein AM study was used to assess the cell wall integrity and the double staining to assess the cytotoxicity showed specific features. The control group of cells was not only brilliantly green but also showed homogeneous spindle shape, indicating the integrity of cell wall and the metabolic potential. The addition of PPF to the basic

Samples	ALP (μg/mL)			
	0.1	1	10	100
AgFp	156.2	132.3	148.3	140.5
ZnFp	136.8	139.6	125.5	129.8
MgFp	143.7	143.4	148.0	138.3
Control	130.3			

Table 1.
Alkaline phosphatase activity of metal-doped fluorophosphates nano bioglass powders.

Samples	Intracellular (Concentration (μg/mL))			Extracellular (Concentration (μg/mL))		
	1	10	100	1	10	100
AgFp	0.091	0.084	0.207	0.319	0.194	0.274
ZnFp	0.011	0.251	0.000	0.208	0.557	0.324
MgFp	0.091	0.084	0.207	0.469	0.300	0.317
Control	0.083	0.083	0.083	0.430	0.430	0.430

Table 2.
Osteocalcin (intracellular and extracellular) activity of metal-doped fluorophosphates nano bioglass powders.

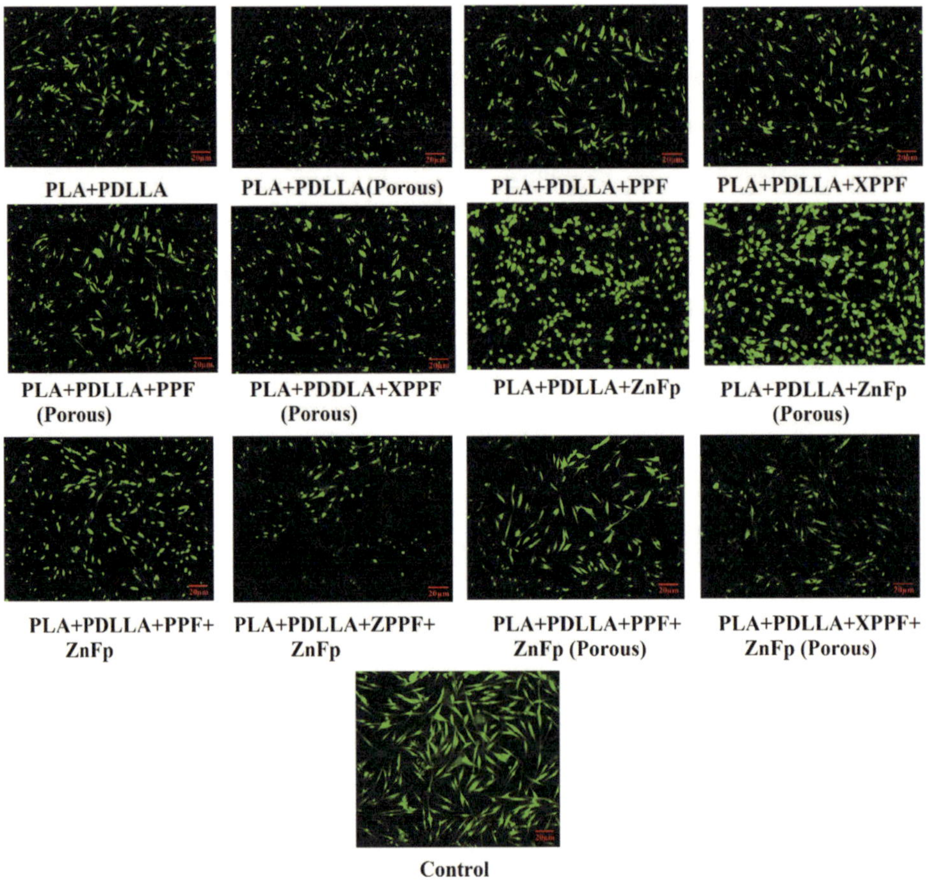

Figure 3.
Calcein AM study of composites.

components PLA+PDLLA increased the cell wall integrity and the addition of pores to the same increased the number of spindle shaped cells [22].

The addition of FP glass to the basic components PLA+PDLLA [23] with pores or without pores increased the number of cells phenomenally, but the quality of them was poor exhibited by their round shape rather than the spindle shape of the healthy cell. When all the components PLA+PDLLA+PPF and ZnFp glasses were added, both the intensity of fluorescence and the quality of the cells also increased, and it was more so when pores were present in the composite (**Figure 3**).The XPPF (cross-linked PPF) when replaced the PPF in the composite, there was only deleterious effect both in the fluorescence and the quality of the cells (**Table 3**).

From the above study, it can be inferred that the least toxic composite was that of PLA+PDLLA+PPF+ZnFp glass.

The three essential gene markers in the synthesis of bone from the stage of mesenchymal stem cells to that of the osteocyte maturation are Osteocalcin [24], Collagen II [25], and RUNX2 [26](**Table 4**). When the results were charted to scrutinize the fold change than the control, the fold increase in collagen II was highest with PLA+PDLLA+PPF+AgFp, and the highest fold increase in osteocalcin was also with AgFp but when constituted with 1,2-diol than with PPF. The highest fold change in

Sample name	Mean	SD	% Cytotoxicity
PLA+PDLLA	50.60	0.19	49.40
PLA+PDLLA(porous)	48.21	1.21	51.79
PLA+PDLLA+PPF	43.70	0.78	56.30
PLA+PDLLA+XPPF	59.18	1.79	40.82
PLA+PDLLA+PPF+ZnFp	87.11	1.20	42.89
PLA+PDLLA+XPPF+ZnFp	42.70	1.36	57.30
PLA+PDLLA+PPF(Porous)	51.79	1.21	48.21
PLA+PDLLA+XPPF(Porous)	60.69	1.51	39.31
PLA+PDLLA+PPF+ZnFp(Porous)	78.37	3.92	21.63
PLA+PDLLA+XPPF+ZnFp(Porous)	69.53	1.71	30.47
PLA+PDLLA+ZnFp	54.23	0.47	45.77
PLA+PDLLA+ZnFp(Porous)	37.99	1.64	62.01

Table 3.
Cytotoxicity of the prepared composites.

Composites	MTT	ALP (IU/mL)	CollagenII (Fold increase)	Osteocalcin (Fold increase)	RUNX2 (Fold increase)	Chondroitin levels (ng/mL)
PLA+1,2-diol+AgFp	90.95	0.443	0.16	2.30	0.04	7.450
PLA+PDLLA+1,2-diol+AgFp	79.84	1.915	0.81	4.70	0.55	11.540
PLA+PPF+AgFp	81.48	0.596	0.40	0.22	0.03	3.250
PLA+PDLLA+PPF+AgFp	81.13	1.406	4.40	3.20	0.50	5.430
PLA+1,2-diol+ZnFp	89.72	0.938	0.12	0.07	0.01	10.210
PLA+PDLLA+1,2-diol+ZnFp	84.36	2.306	1.70	1.20	0.40	3.850
PLA+PPF+ZnFp	93.42	0.503	0.50	1.40	0.09	13.230
PLA+PDLLA+PPF+ZnFp	87.20	1.543	1.10	2.60	0.98	9.780
PLA+1,2-diol+MgFp	90.54	1.081	0.25	0.05	0.78	10.880
PLA+PDLLA+1,2-diol MgFp	81.48	1.208	0.93	0.19	0.78	8.210
PLA+PPF+MgFp	53.09	1.39	0.23	0.23	0.21	5.430
PLA+PDLLA+PPF+MgFp	86.42	1.544	0.53	0.24	0.20	10.210
Control	100	1.529	—	—	—	0.180

Table 4.
MT, ALP, collagen II, osteocalcin, RUNX2, and chondroitin sulfate levels of composites.

RUNX2 than the control was with ZnFp when combined with PLA+PDLLA+PPF. All the Mg-based composites fared poorly with all the three types of gene markers (**Figure 4** and **Table 4**).

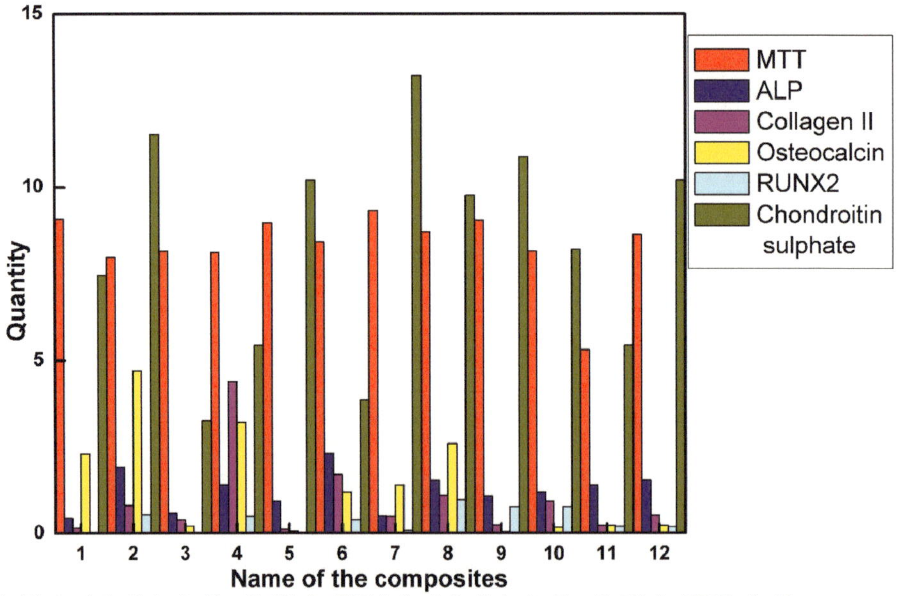

1. PLA+1,2-diol+AgFp, 2. PLA+PDLLA+1,2-diol+AgFp, 3. PLA+PPF+AgFp, 4. PLA+PDLLA+PPF+AgFp, 5. PLA+1,2-diol+ZnFp, 6. PLA+PDLLA+1,2-diol+ZnFp, 7. PLA+PPF+ZnFp, 8. PLA+PDLLA+PPF+ZnFp, 9. PLA+1,2-diol+MgFp, 10.PLA+PDLLA+1,2-diol+MgFp, 11. PLA+PPF+MgFp, 12. PLA+PDLLA+PPF+MgFp.

Figure 4.
Pie chart of the MTT,ALP, collagen II, osteocalcin, RUNX2, and chondroitin sulfate levels of composites of varying composition.

The results showed that all the 12 composites showed many fold increase in the secretion of chondroitin sulfate [27] than the control, immaterial of the component having 1,2-diol or PPF and the FPglass being either Ag, Zn, or Mg.

The pH variation of all the compression-molded specimens showed uniformly a reduction in the first 2 days, which is because of phosphoric acid formation. And all the specimens bounced back to 7 on the third day due to the alkaline earth metal (Na^+ and/or Ca^{2+}) release. The dissolution of the ions thus replaces H^+ ions by cations (Na^+ and/or Ca^{2+}) leading to an increase in hydroxyl ion concentration. None of them went below 6.5 even in the first 2 days. From then on it showed a steady variation between 7 and 6.7.

The strip and cylinders made by gel foam casting under rapid heating showed a better pH even in the first 2 days and never went below 6.8, and the end stage also showed higher pH than the scaffolds. The highest pH reached was with the strip of scaffold made by rapid heating method, and it was 7.15. This variation shows the better homogenezity and the porosity achieved by the rapid heating method, which avoids high acidic environment that can lead onto graft rejection [28]. The crystal formation over the strip and cylinder after *in vitro* evaluation is shown in **Figure 5**.

The predominant functional groups present in the composites were studied using their respective FTIR spectra:alcohol (3200–3500 cm^{-1}), alkanes (2850–3000 cm^{-1}), saturated ketone (1735–1750 cm^{-1}), alkenes (1630–1680 cm^{-1}), asymmetric methyl bend (1450–1470 cm^{-1}), and methyl bending (1350–1395 cm^{-1}). The presence of P-O bend (560–500 cm^{-1}) bands indicates the formation of calcium phosphate($CaO-P_2O_5$) layer. The carbonate group $(CO_3)^{2-}$ (1400–1550 cm^{-1}) bands showed the crystalline

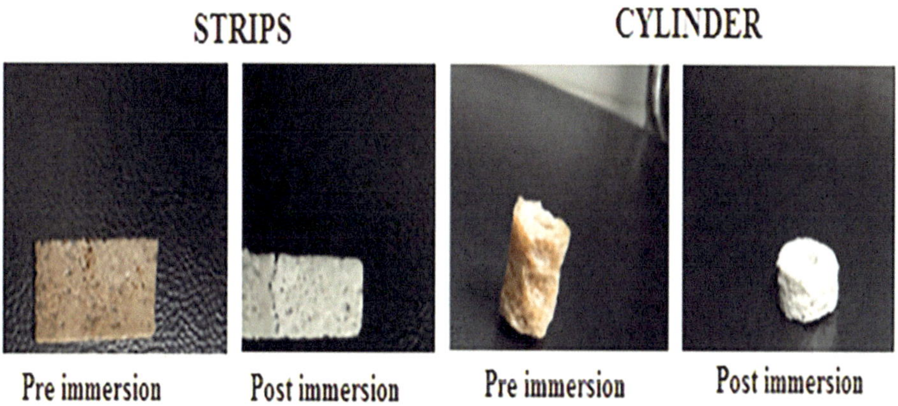

Figure 5.
Pre- and post-immersion photographs of the strip and cylinder.

nature of the HAp layer. The bands are observed at above 3500 cm^{-1}, which corresponds to the OH group. After 21 days of soaking in SBF, the strong intensity and frequency shift of the $(CO_3)^{2-}$,P-O-P stretch and P-O bend groups reveal the interaction of the composite and HAp precipitation [29].

The shoulder peak at 1450–1410 cm^{-1} coupled with the weaker peak at 870–875 cm^{-1} corresponds to type B carbonate vibrations, whereas the vibration regions of type A carbonate are 1450–1410 cm^{-1} coupled with a band at 880 cm^{-1}. The type A and B carbonates [30, 31] are indistinguishable in these scaffolds because the ester peaks also lie on the same region. Both type A and B carbonates are present in these scaffolds and their intensities are maximum at three selected scaffold composites (PLA+PDLLA+PPF+ZnFp, PLA+PDLLA+PPF+AgFp, PLA+PDLLA+PPF+MgFp), For the same composites, the corresponding peaks for HAp in rapid heating combined gel foam casting are higher than the compression-molded scaffolds.

Although the HAp precipitation was noted in all the fabricated scaffolds, the intensity of the carbonated group $(CO_3)^{2-}$and phosphatebased group (P-O-P asymmetric and symmetric stretch, P-O bend) was observed as high in gel foam casting under rapid heating.

The SEM evaluation of the scaffolds and strips was done after gold sputtering. (Model Ultra 55; Zeiss, Oberkochen, Germany). After evaluating the surface apatite formation, the specimens were cut into two halves and turned by 90°, and the depth of the apatite formation was measured. There was no significant change in the percentage, and it was infered that all the composites have near equal conversion once the pores allow penetration of the SBF inside except the absence of PDLLA had negative significance in the extent of crystalline conversion (**Table 5**).

The similar specimens were subjected to *in vitro* evaluations, which were analyzed by the same way in the same Scanning Electron Microscope to assess the degree of surface pores and the change in crystallinity after *in vitro* study. The photograph of a stirp of composite and a cylindrical composite, both made by gel foam casting under rapid heating, shows the retention of the shape after SBF immersion for 21 days, but there was complete change in the color and the texture indicating the crystalline conversion. The SEM of scaffold in two different magnifications both before and after *in vitro* evaluation is shown in **Figure 6**, which shows very scarce

Composites	Crystallinity(%)
PLA+1,2-diol+AgFp	86.88
PLA+PDLLA+1,2-diol+AgFp	97.42
PLA+PPF+AgFp	49.36
PLA+PDLLA+PPF+AgFp	72.70
PLA+1,2-diol+ZnFp	59.54
PLA+PDLLA+1,2-diol+ZnFp	78.89
PLA+PPF+ZnFp	87.31
PLA+PDLLA+PPF+ZnFp	76.19
PLA+1,2-diol+MgFp	73.33
PLA+PDLLA+1,2-diol MgFp	78.68
PLA+PPF+MgFp	94.62
PLA+PDLLA+PPF+MgFp	82.09
PLA+PDDLA+PPF+AgFp(Strip)	73.62
PLA+PDDLA+PPF+ZnFp (Strip)	81.42
PLA+PDDLA+PPF+MgFp(Strip)	87.34

Table 5.
Apatite conversion(%).

amount of crystallization in the pre *in vitro* evaluation and the homogeneous pores being well exhibited. After 21 days of immersion in SBF, the crystalline conversion is observed, and all the pores have been near completely clogged with the crystals formed [32, 33].

The SEM micrographs of the scaffolds (pre- and post-immersion). In pre-immersion status shows specks of crystallization indicating the high hydrophilicity of the scaffold, and the post-immersion evaluation of the same shows complete formation to crystals, which proves the high bioresorbability of the scaffold. The SEM evaluation of multilayered scaffold made by rapid heating under low magnification shows the adequacy of pores. The pre-immersion and the post-immersion SEM micrographs clearly show the formation of sufficient crystals. The EDAX evaluation of the pre and post in *vitro* SEM confirms the high level of carbonated hydroxyapatite and fluroapatite formation in the scaffold.

The Micro-CT (**Figure 7**) evaluation of the cylindrical scaffold made by gel foam casting under rapid heating proved the following factors: a) The scaffold had no layering and was continuous. b) There was adequate porosity and the pore sizes were varying. c) The pores were all well connected by interpores. The same specimen after *in vitro* evaluations had sufficient fomation of crystals with the preservation of the deeper pores [34].

4.2 *In vivo* studies

From the *in vitro* evaluation, PPF-based composites were shortlisted for *in vivo* studies.

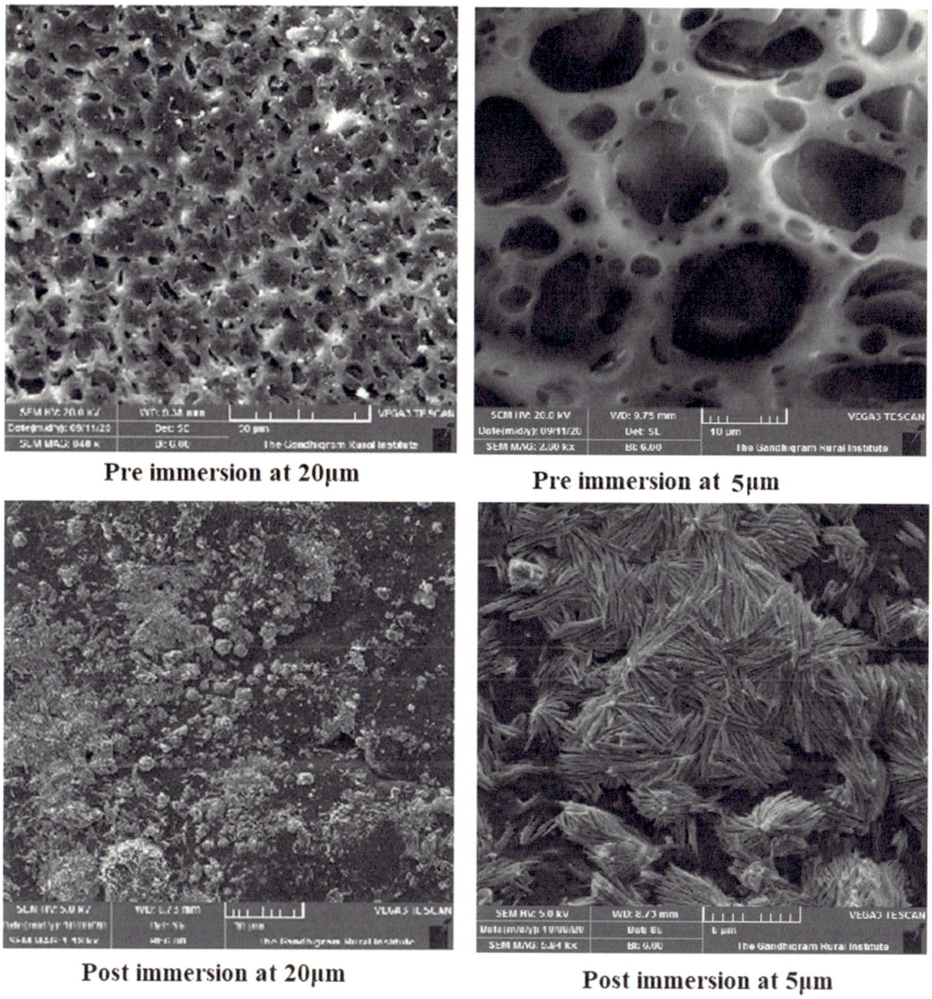

Figure 6.
Pre- and post-immersion of the scaffolds.

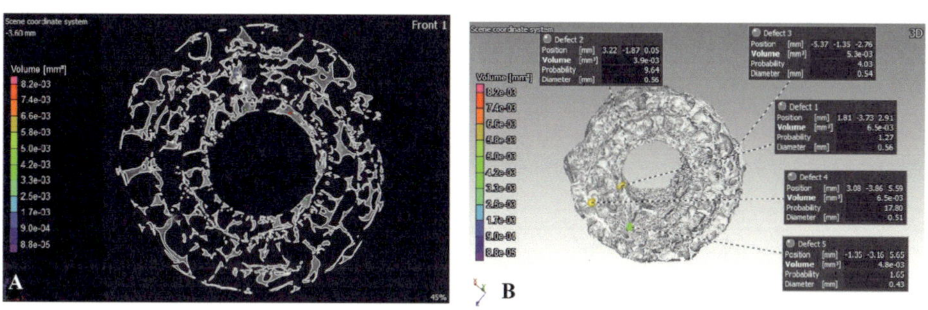

Figure 7.
Micro-CT studies of the cylindrical composite. (A - Pre-immersion and B – Post-immersion).

4.2.1 Granules

It was found that the femur had fractured and the ends were apart (**Figure 8**). Neither immobilization of the femur or any form of fixation was done. The rabbit was not limping and was feeding well. There was only a flare of the ends of the fracture and there was no evidence of any callus on the ninth postoperative day. After another week (Day 16), the limb when examined clinically had sound union. The X-ray taken showed abundant callus not only in the fracture end but all along the femur where the trough had been made and even below.

The animal was euthanized, the limb harvested, skin and muscles were peeled off, and an abundant amount of callus was found to have united the fracture very strongly. The dissected specimen were studied by X-ray and the specimen were preserved in 10% formalin.

The specimen was prepared and the decalcified specimen was sectioned axially to exhibit the two segments of the femur with the intervening tissue formed. The

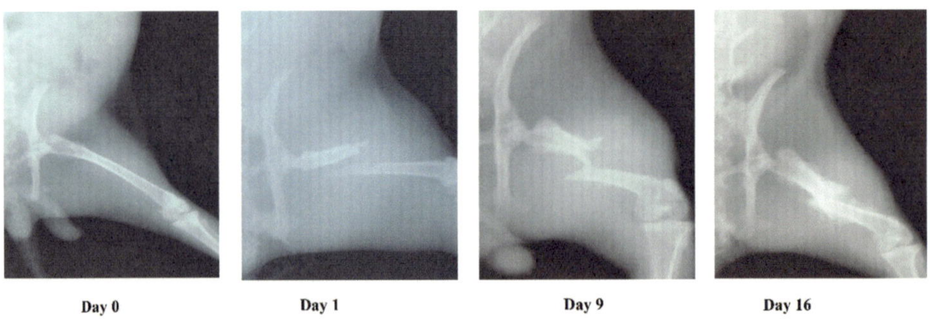

Figure 8.
X-ray photographs of the opearted site in rabbit.

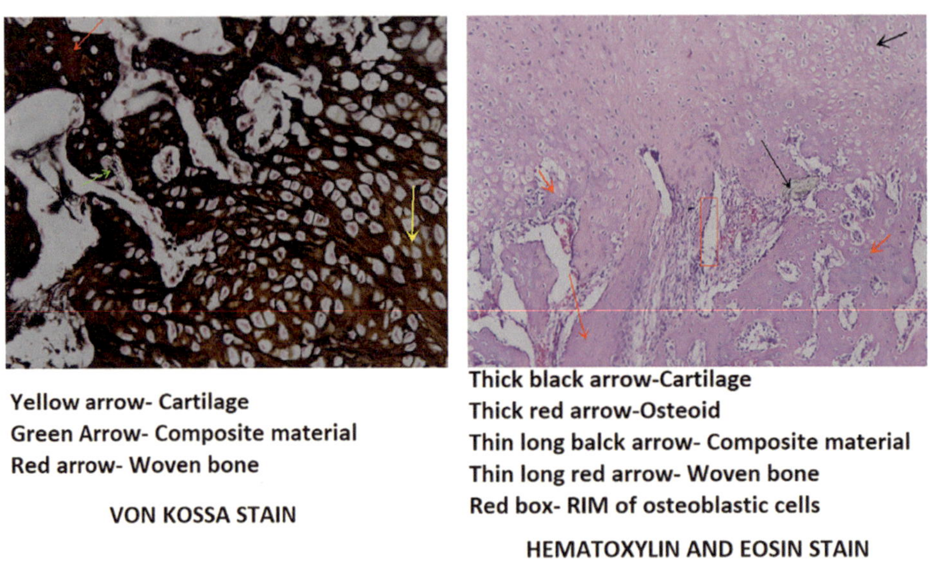

Figure 9.
Histopathological study of the dissected specimen.

specimen was stained using regular eosin-hematoxylin stain and also von kossa stain. (**Figure 9**).

The procedure done using granules of the composite has been serially shown in the photographs (**Figure 1**). Though the limb got fractured, it did not receive any specific treatment for it. Until ninth day, there was a scarce response to heal but by the 16th day it had soundly united. (**Figure 8**). The X-ray of the specimen after dissection showed the extension of the callus almost over the entire femur. The histopathological evaluation was specifically focused toward the tissue between the fractured ends where the granules had been packed. The significant observations were: a) Nearly the whole of the granules had resorbed except occasional trace of it. b) Abundant cartilage had formed between the ends indicating the enchondral ossification. c) Woven bone formed in between the ends of the fracture was a proof of the rapidity of the fusion occurring. d) The absence of multinucleated giant cells indicates the biocompatibility of the composite. e) Similar features were observed in both the staining (**Figure 9**). The modified tetrachrome staining throws much more information than the above two. a) The new lamellar bone formed in continuity with the resorbing composite granule. b) The sound union by the woven bone formed from chondral ossification. c) The abundance of osteoblasts and the osteoid. d) An exuberant neo vascularization among the fibroblasts is well seen (**Figure 10a–d**) [35, 36].

4.2.2 Strip

The X-ray photographs show no evidence of the placed composite strip or the corticotomy made as the composite is not radio opaque and the furrow is very narrow. But the X-rays taken on the ninth day showed all three animals had fractured their femur. No specific treatment such as immobilization or interference was done for the fracture. Clinical union occurred as early as 15th day and was confirmed by X ray on 16th day (**Figure 11**) and CT scan on 19th day (**Figure 12**). The harvested limb after euthanizing

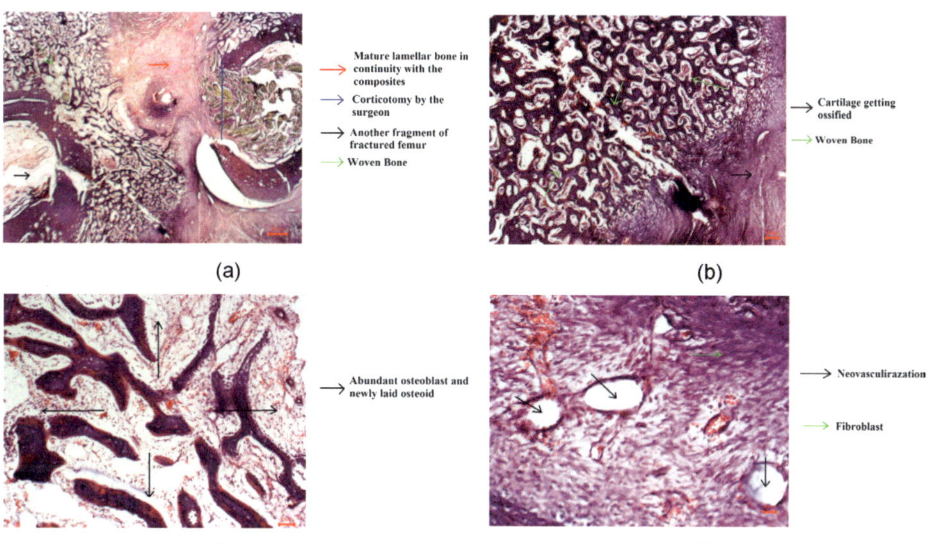

Figure 10.
a-d represent the tetrachrome stain results of the granules (Composition: PLA+PDLLA+PPF+AgFp).

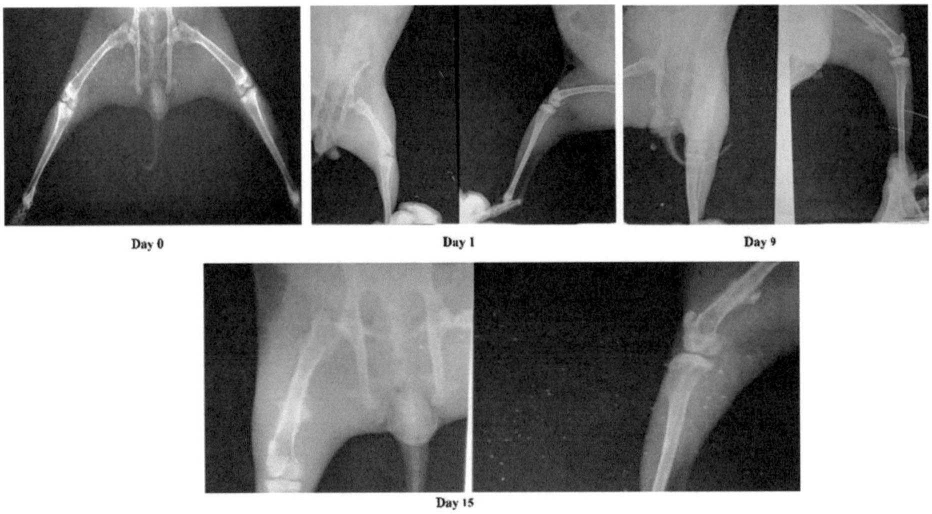

Figure 11.
Studies using composite strips: X-ray photographs of the operated site in rabbit.

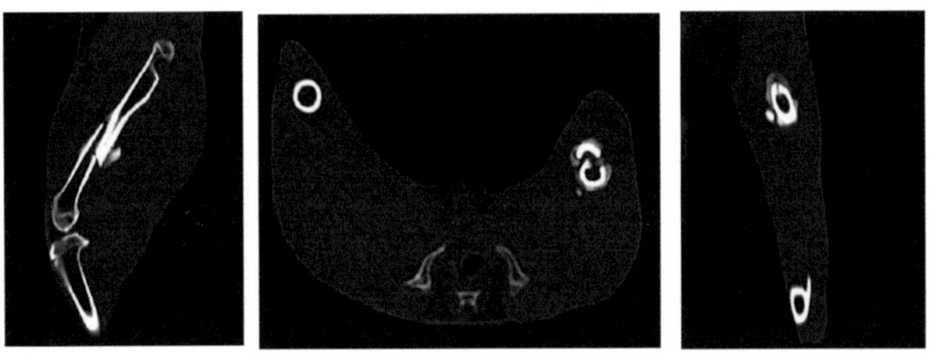

Figure 12.
Studies using composite strips: CT scan of the fractured leg in rabbit.

the animal showed the composite strip was adherent to the bone underneath. The X-ray of the specimens showed abundant callus along the fracture (**Figure 13**) and the composite strip was not seen in the X-ray.

The histopathological evaluation showed the following features: a) Both the layers of the scaffold had merged into one layer. b) The composite had attached to the bone beneath. c) There was abundant woven bone formed beneath the composite strip at the level of the corticotomy. d) The second layer of the composite strip kept away from the corticotomy had profuse infiltration of fibrocytes. e) The fibrous change-over in the superficial layer of the composite had abundant neovascularization These changes confirm the osteo induction potential of the composite, the ability of the composite to go for bioconversion, and high bioactivity of the composite [37–39].

The modified tetrachrome staining of the specimens with the cross section at the level of the composite confirmed the findings by EH stain and showed the additional features. **Figure 14a** shows conversion of the fragmented composite forming woven bone to heal the corticotomy made and the binding of the two layers of the composite

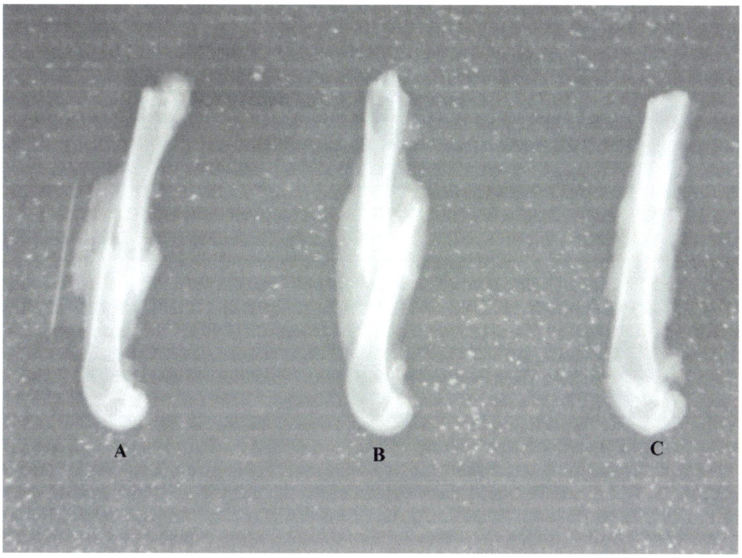

Figure 13.
Studies using composite strips: Postoperative X-ray of the dissected specimen from rabbits. (A-PLA+PDLLA+PPF+AgFp; B-PLA+PDLLA+PPF+ZnFp; C-PLA+PDLLA+PPF+MgFp).

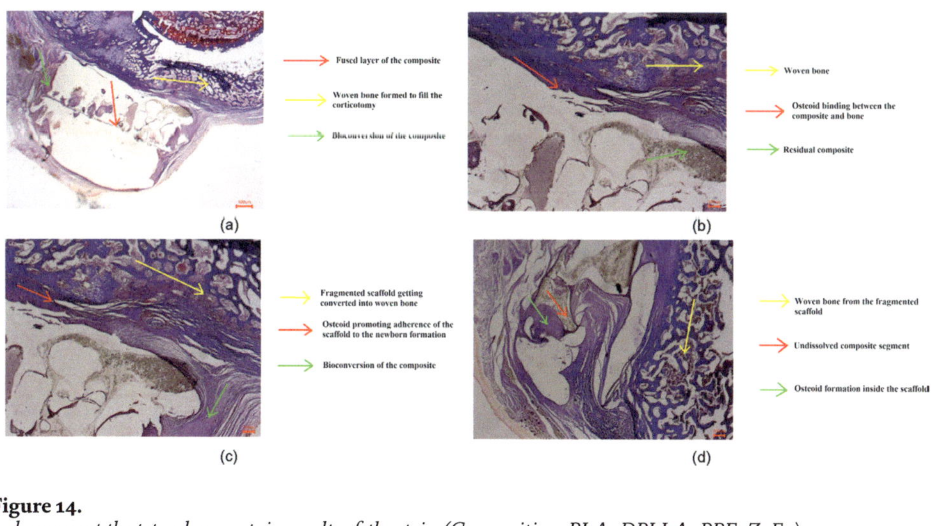

Figure 14.
a-d represent the tetrachrome stain results of the strip (Composition: PLA+DPLLA+PPF+ZnFp).

strip and random infiltration of the layer close to the bone with fibroblasts and specks of osteoid. On higher magnification (**Figure 14b**), the fusion of the composite strip to the underlying bone by osteoid is well seen. On further magnification (**Figure 14c**), the infiltration of the composite by newly formed layers of osteoid is well made out replacing the dissolved area of the composite. **Figure 14d** shows the adhesion of the composite strip, the composite strip dissolving and disintegrating to form new woven bone healing the corticotomy, the abundant laying of new osteoid in the dissolved portion of the composite.

5. Conclusions

The novel composite of PLA+PDLLA+PPF+AgFp/ZnFp/MgFp meets most of the requirements of an ideal bone substitute, and it bridges the gap between the need of the clinician and the biomaterial scientist, more than the available present-day commercial ceramic composites. It is not only conductive like HAp but also inductive. It is more inductive than Tri calcium phosphate. It takes less time for resorption than Bioglass and the presence of fluoride makes the rate of bioconversion high. As the end products of all the four components of the composite are natural ingredients of the body and induce bone formation by enhancing the genetic pathway, the composite is a real biomimetics. Though small animal studies have proved the usefulness of the composite, their efficacy has to be confirmed in clinical situations in large animals before human trial.

Conflict of interest

The authors declare no conflict of interest.

Author details

Pugalanthipandian Sankaralingam*, Poornimadevi Sakthivel and Vijayakumar Chinnaswamy Thangavel
Bone Substitutes, Madurai, India

*Address all correspondence to: drpandian@bonesubstitutes.in

References

[1] Calori GM, Mazza E, Colombo M, Ripamonti C. The use of bone-graft substitutes in large bone defects: Any specific needs? Injury. 2011;**1**(42):S56-S63. DOI: 10.1016/j.injury.2011.06.011

[2] Fernandez de Grado G, Keller L, Idoux-Gillet Y, Wagner Q, Musset AM, Benkirane-Jessel N, et al. Bone substitutes: A review of their characteristics, clinical use, and perspectives for large bone defects management. Journal of Tissue Engineering. 2018;**2**(9):6819

[3] Thirumalai J. Introductory chapter: The testament of hydroxyapatite: New prospects in regenerative medicinal treatments. Hydroxyapatite. 2018;**14**: 3-14. DOI: 10.5772/intechopen.72767

[4] Apelt D, Theiss F, El-Warrak AO, Zlinszky K, Bettschart-Wolfisberger R, Bohner M, et al. *In vivo* behavior of three different injectable hydraulic calcium phosphate cements. Biomaterials. 2004;**25**(7-8):1439-1451. DOI: 10.1016/j.biomaterials.2003.08.073

[5] Theiss F, Apelt D, Brand B, Kutter A, Zlinszky K, Bohner M, et al. Biocompatibility and resorption of a brushite calcium phosphate cement. Biomaterials. 2005;**26**(21):4383-4394. DOI: 10.1016/j.biomaterials.2004.11.056

[6] Daculsi G, Laboux O, Malard O, Weiss P. Current state of the art of biphasic calcium phosphate bioceramics. Journal of Materials Science. Materials in Medicine. 2003;**14**(3):195-200. DOI: 10.1023/a:1022842404495

[7] Grgurevic L, Pecina M, Vukicevic S, Marshall R. Urist and the discovery of bone morphogenetic proteins. International Orthopaedics. 2017;**41**(5):1065-1069. DOI: 10.1007/s00264-017-3402-9

[8] Hench LL, Splinter RJ, Allen WC, Greenlee TK. Bonding mechanisms at the interface of ceramic prosthetic materials. Journal of Biomedical Materials Research. 1971;**5**(6):117-141. DOI: 10.1002/jbm.820050611

[9] LeGeros RZ, Silverstone LM, Daculsi G, Kerebel LM. *In vitro* caries-like lesion formation in F^-containing tooth enamel. Journal of Dental Research. 1983;**62**(2):138-144. DOI: 10.1177/00220345830620021101

[10] Knowles JC. Phosphate based glasses for biomedical applications. Journal of Materials Chemistry. 2003;**13**(10):2395-2401. DOI: 10.1039/B307119G

[11] Gurusamy R, Sankaralingam P. Inventors; Pandian Bio Medical Research Centre, assignee. Bioactivity of Fluorophosphate Glasses and Method of Making Thereof. 2015 Jun 18.WO2015087345A1

[12] Kasper FK, Tanahashi K, Fisher JP, Mikos AG. Synthesis of poly (propylene fumarate). Nature Protocols. 2009;**4**(4):518-525. DOI: 10.1038/nprot.2009.24

[13] Siva Kaylasa Sundari S, Shamim Rishwana S, Poornimadevi S, Vijayakumar CT. Synthesis of macromolecular brush and its thermal degradation studies. International Journal of Polymer Analysis and Characterization. 2022;**27**(3):147-157. DOI: 10.1080/1023666X.2022.2029263

[14] Kashif I, Soliman AA, Sakr EM, Ratep A. Effect of different conventional melt quenching technique on purity of

lithium niobate ($LiNbO_3$) nano crystal phase formed in lithium borate glass. Results in Physics. 2012;**2**:207-211. DOI: 10.1016/j.rinp.2012.10.003

[15] Rodrigues N, Benning M, Ferreira AM, Dixon L, Dalgarno K. Manufacture and characterisation of porous PLA scaffolds. Procedia Cirp. 2016;**49**:33-38. DOI: 10.1016/j.procir.2015.07.025

[16] Khang G, Kim MS, Lee HB. A manual for biomaterials/scaffold fabrication technology. World Scientific Publishing Company Singapore; 2007 Jul 3. https://doi.org/10.1142/6408. ISBN: 978-981-3101-60-9

[17] Bejarano J, Detsch R, Boccaccini AR, Palza H. PDLLA scaffolds with Cu-and Zn-doped bioactive glasses having multifunctional properties for bone regeneration. Journal of Biomedical Materials Research Part A. 2017;**105**(3):746-756. DOI: 10.1002/jbm.a.35952

[18] Oh SA, Kim SH, Won JE, Kim JJ, Shin US, Kim HW. Effects on growth and osteogenic differentiation of mesenchymal stem cells by the zinc-added sol-gel bioactive glass granules. Journal of tissue engineering. 2010;**1**(1):475260. DOI: 10.4061/2010/475260

[19] Trivedi S, Srivastava K, Gupta A, Saluja TS, Kumar S, Mehrotra D, et al. A quantitative method to determine osteogenic differentiation aptness of scaffold. Journal of Oral Biology and Craniofacial Research. 2020;**10**(2):158-160. DOI: 10.1016/j.jobcr.2020.04.006

[20] Florencio-Silva R, Sasso GR, Sasso-Cerri E, Simões MJ, Cerri PS. Biology of bone tissue: Structure, function, and factors that influence bone cells. Bio Med Research International 2015;**2015**:17. Article ID: 421746. https://doi.org/10.1155/2015/421746

[21] Zoch ML, Clemens TL, Riddle RC. New insights into the biology of osteocalcin. Bone. 2016;**82**:42-49. DOI: 10.1016/j.bone.2015.05.046

[22] Cecen B, Kozaci D, Yuksel M, Erdemli D, Bagriyanik A, Havitcioglu H. Biocompatibility of MG-63 cells on collagen,poly-L-lacticacid,hydroxyapatite scaffolds with different parameters. Journal of Applied Biomaterials and Functional Materials. 2015;**13**(1):10-16. DOI: 10.5301/jabfm.5000182

[23] Ronca A, Ambrosio L, Grijpma DW. Design of porous three-dimensional PDLLA/nano-hap composite scaffolds using stereolithography. Journal of Applied Biomaterials and Functional Materials. 2012;**10**(3):249-258. DOI: 10.5301/JABFM.2012.10211

[24] Meretoja VV, Tirri T, Malin M, Seppälä JV, Närhi TO. Enhanced osteogenicity of bioactive composites with biomimetic treatment. BioMed Research International. 2014;**2014**:207676. pp. 1-8 DOI: 10.1155/2014/2076762014:207676

[25] Chiu LH, Lai WF, Chang SF, Wong CC, Fan CY, Fang CL, et al. The effect of type II collagen on MSC osteogenic differentiation and bone defect repair. Biomaterials. 2014;**35**(9):2680-2691. DOI: 10.1016/j.biomaterials.2013.12.005

[26] Dong M, Jiao G, Liu H, Wu W, Li S, Wang Q, et al. Biological silicon stimulates collagen type 1 and osteocalcin synthesis in human osteoblast-like cells through the BMP-2/Smad/RUNX2 signaling pathway. Biological Trace Element Research. 2016;**173**(2):306-315. DOI: 10.1007/s12011-016-0686-3

[27] Wibowo H, Widiyanti P, Asmiragani S. The role of chondroitin sulfate to bone healing indicators and compressive strength. Journal

of Basic and Clinical Physiology and Pharmacology. 2021;**32**(4):631-635. DOI: 10.1515/jbcpp-2020-0406

[28] Sankaralingam P, Sakthivel P, Subbiah A, Periyasamy A, Begam J, Thangavel C. Fluorophosphate bio-glass for bone tissue engineering: *in vitro* and *in vivo* study. Bioinspired, Biomimetic and Nanobiomaterials. 2021;**10**(4):123-131. DOI: 10.1680/jbibn.21.00025

[29] Beherei HH, Mohamed KR, El-Bassyouni GT. Fabrication and characterization of bioactive glass (45S5)/titania biocomposites. Ceramics International. 2009;**35**(5):1991-1997

[30] Ren FZ, Leng Y. Carbonated apatite, type-A or type-B?. In: Key Engineering Materials. Switzerland. Trans Tech Publications Ltd. 2012;**493**:293-297

[31] Hanifi A, Fathi MH. Bioresorbability evaluation of hydroxyapatite nanopowders in a stimulated body fluid medium. Iranian Journal of Pharmaceutical Sciences. 2008;**4**(2):141-148

[32] Xu Z, Hodgson MA, Cao P. Effect of immersion in simulated body fluid on the mechanical properties and biocompatibility of sintered Fe–Mn-based alloys. Meta. 2016;**6**(12):309

[33] Ofudje EA, Adeogun AI, Idowu MA, Kareem SO. Synthesis and characterization of Zn-doped hydroxyapatite: Scaffold application, antibacterial and bioactivity studies. Heliyon. 2019;**5**(5):e01716

[34] Palmroth A, Pitkänen S, Hannula M, Paakinaho K, Hyttinen J, Miettinen S, et al. Evaluation of scaffold microstructure and comparison of cell seeding methods using micro-computed tomography-based tools. Journal of the Royal Society, Interface. 2020;**17**(165):102

[35] Ralis ZA, Watkins G. Modified tetrachrome method for osteoid and defectively mineralized bone in paraffin sections. Biotechnic & Histochemistry. 1992;**67**(6):339-345. DOI: 10.3109/10520299209110046

[36] Chen X, Zhao Y, Geng S, Miron RJ, Zhang Q, Wu C, et al. *In vivo* experimental study on bone regeneration in critical bone defects using PIB nanogels/boron-containing mesoporous bioactive glass composite scaffold. International Journal of Nanomedicine. 2015;**10**:839. DOI: 10.2147/IJN.S69001

[37] Rentsch C, Schneiders W, Manthey S, Rentsch B, Rammelt S. Comprehensive histological evaluation of bone implants. Biomatter. 2014;**4**(1):e27993. DOI: 10.4161/biom.27993

[38] Hsieh MK, Wu CJ, Su XC, Chen YC, Tsai TT, Niu CC, et al. Bone regeneration in Ds-red pig calvarial defect using allogenic transplantation of EGFP-pMSCs–A comparison of host cells and seeding cells in the scaffold. Plo S One. 2019;**14**(7):e0215499. DOI:10.1371/journal.pone.0215499

[39] Sankaralingam P. Inventors; BoneSubstitutes, assignee. A synthetic composite as bone graft and the method thereof. 2021 Sep 02. WO/2021/171315

Section 3

Biomimetics at the Nanoscale

Chapter 5

Bio-Simulation of the Induction of Forced Resonance Mechanical Oscillations to Virus Particles by Non-Ionizing Electromagnetic Radiation: Prospects as an Anti-Virus Modality

Nikolaos K. Uzunoglu

Abstract

The induction of acoustic-mechanical oscillations to virus particles by illuminating them with microwave signals is analyzed theoretically. Assuming the virus particle is of spherical shape, its capsid consisting primarily of glycoproteins, a viscous fluid model is adopted while the outside medium of the sphere is taken to be the ideal fluid. The electrical charge distribution of virus particles is assumed to be spherically symmetric with a variation along the radius. The generated acoustic-mechanical oscillations are computed by solving a boundary value problem analytically, making use of Green's function approach. Resonance conditions to achieve maximum energy transfer from microwave radiation to acoustic oscillation to the particle are investigated. Estimation of the feasibility of the technique to compete with virus epidemics either for sterilization of spaces or for future therapeutic applications is examined briefly.

Keywords: virus mechanics, acoustic radiation biomedical effects

1. Introduction

The study of the physical properties of various types of viruses has attracted significant interest from several interdisciplinary research groups during the last 10 years [1]. Mechanical properties of virus particles with diameters of 10–300 nm, in particular the capsids enclosing the virus gene structures (DNA, RNA both single- and double-stranded), have been studied experimentally using atomic force and electron microscopy (AFM) [2, 3]. Also, elasticity theory methods are applied to conclude the mechanical properties of virus particles [4]. Electric charge distributions of the virus have also been studied by several researchers [5–9]. It is observed that under physiological conditions of salinity and acidity, virus capsid assembly requires the presence of genomic material that is oppositely charged to the core proteins [10]. Furthermore, few researchers have focused their research on the possibility of inducing photon-phonon interactions [11] in virions, which are the virus causing infections [12–17]. Already resonance phenomena of the H3N2 and H1N1 viruses have been demonstrated experimentally [18] , leading to a high rate extinction of them at a resonance microwave frequency near 8 GHz. The physical phenomenon attributed to this interaction is the separation of positive-negative electric charges on the body of the virus particles and the coupling of microwave energy through the interaction with the three-dimensional bipolar electric charge distributions, generating mechanical oscillations at the same frequency. At specific microwave frequencies depending on the diameter and other properties of the particle [19, 20], primarily the dipole acoustic mode, have been claimed and strong coupling leading to high level virus killing rates have been demonstrated recently [19, 20]. The effects of hydration levels on the bandwidth of microwave resonant absorption induced by confined acoustic vibrations have also been studied [21]. It should be stated that the involved phenomenon is of non-thermal nature related to non-ionizing radiation, in this case being the I band microwaves (6–10 GHz). Raman scattering phenomena [22] have also verified the existence of acoustic-mechanical resonance phenomena in virus particles [23].

The recent ongoing Covid-19 worldwide pandemic [24] and its severe consequences make it attractive to investigate the possibility of utilizing the above-mentioned resonance phenomenon either in sterilization of spaces [25] such as clinics, public venues, hospitals or in the future as a therapeutic modality in some cases. In this direction, the possibility of utilizing similar methods used in microwave-induced hyperthermia, to raise the temperature of malignant tumors inside the human body, could be envisaged as a therapeutic modality. In fact, contrary to hyperthermia where usually lower frequencies of 27–2450 MHz are used [26], in this case much higher frequencies (6–8 GHz) were needed to be used. In some cases, ultrasound and laser radiation modalities have also been used, in clinical hyperthermia, along the low-frequency microwave radiation using endo-cavitary radiators. However, in the present case, the interaction of non-ionizing radiation with tissues will be entirely different from hyperthermia to compete with the virus populations. The rather high frequencies used in the present resonance phenomenon pose a challenging problem to penetrate with high-intensity electric fields inside the human body such as in the case of the lungs. However, in the case of the larynx and throat and even some parts of the lungs, endo-cavitary radiators [27], as done in hyperthermia, could be used. Finally, since in the present case, the action of microwave radiation has the character of a resonance interaction, it is foreseen not to need longtime irradiation, contrary to hyperthermia, which in order to raise the tissues from 43 to 45°C needs usually 45–60 minutes. This principle allows to propose in present case short duration pulsed-periodic high-intensity microwave signals [28, 29]. This is expected to alleviate to some degree the penetration problem of electromagnetic energy to the human body.

In all the mentioned publications of this resonance phenomenon, the virus particle is assumed of being an elastic particle, as was modeled by H. Lamb in 1887 [30] for the oscillations of an ideal spherical isolated in space. In the present chapter, the mathematical analysis is carried out also considering the surrounding medium of the virus particle and taking into account the interaction of external microwave radiation with the electric charge distribution of the virus particle. Based on the recently published data on the structure [31] of the Covid-19 virion being 100–150 nm in diameter, because of its reach liposome capsid with few proteins on it, the present work leads us to adopt the model of the spherical virus particle as a viscous fluid while the outer space is taken to be an ideal fluid, with different acoustic characteristics of the spherical particle. Furthermore, vortex phenomena in modeling the viscous virus structure are neglected, since it is assumed that these are very weak and they have no effect on the resonance phenomenon to be studied.

The bio-simulation of the forced oscillations of a virus particle is carried out in the following mathematical steps presented with a flow diagram (**Figure 1**).

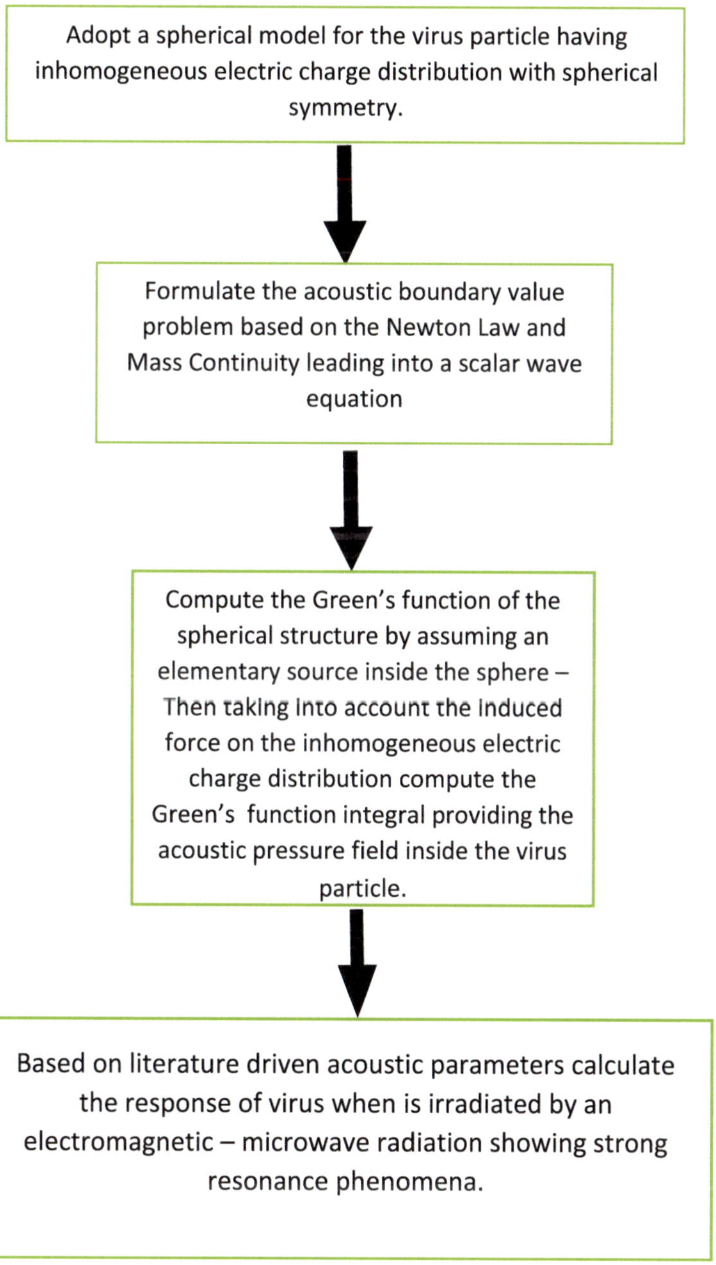

Figure 1.
A flow chart diagram of the bio-simulation analysis of the interaction of virus particles with electromagnetic waves inducing acoustic oscillations inside the virus particle.

2. Mathematical formulation of the phenomenon

A spherical particle of radius α, shown in **Figure 2**, is assumed to pose a continuous electric charge distribution with spherical symmetry defined by the equation

$$\rho_q(\mathbf{r}) = \frac{Q}{\sigma}\left(1 - \frac{5}{3}\left(\frac{r}{\alpha}\right)^2\right) \quad (1)$$

where Q is the total positive electric charge in the center of the sphere, $\sigma = 8\pi\alpha^3 (3/5)^{3/2}/15$ is a normalization constant. The term 5/3 in the

Above Eq. (1) was selected to have the total charge of the particle to be zero, that is to have a balance between the positive (inner $r<\alpha(3/5)^{1/2}$ region) and negative (towards the external surface) charge distributions. It is evident that the particle could have the opposite charge distribution and the same analysis is valid. The proposed method is extendable to the case of non-symmetric charge distribution, and then higher order modes will be excited.

The spherical model of the virus is assumed to be a compressible fluid, characterized by its homogenous mass density ρ_1 acoustic wave propagation speed c_1, and total viscosity constant (dynamic and bulk) χ. Then, assuming $e^{j\omega t}$ as time dependence, the propagation of acoustic wave phenomena is described by the following field equations [32]:

3. Newton law

$$\rho_1 j\omega \boldsymbol{v}_1(\boldsymbol{r}) = -\nabla P_1(\boldsymbol{r}) + x\nabla(\nabla.\boldsymbol{v}_1(\boldsymbol{r})) + \boldsymbol{f} \quad (2)$$

where $\boldsymbol{v}_1$ is the velocity, P_1 the pressure field and $\boldsymbol{f}$ is the force density (N/m^3) because of the electric charge distribution inside the sphere.

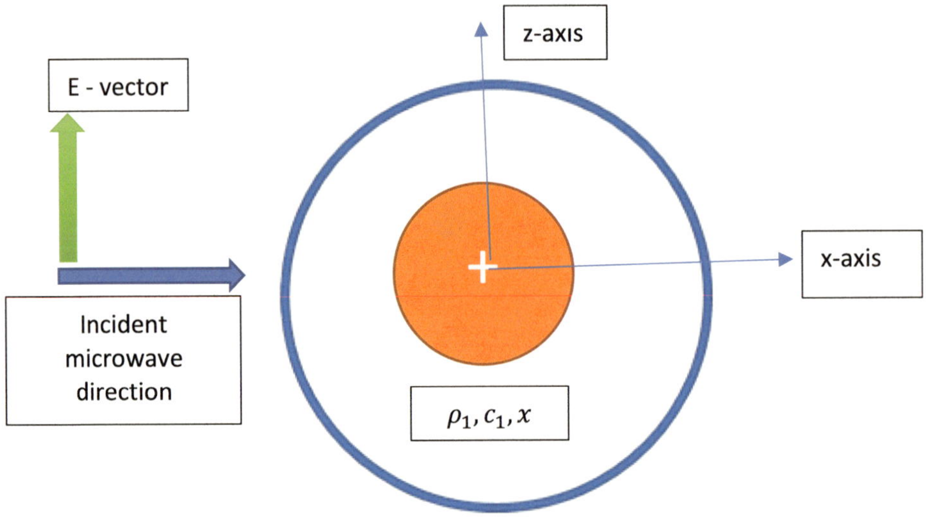

Figure 2.
Spherical model for the virus particle.

4. Mass continuity equation

$$j\omega P_1(\boldsymbol{r}) = -c_1^2\rho_1\nabla.v_1(\boldsymbol{r}) \tag{3}$$

The force density term $\boldsymbol{f}$ in Eq. (2), taking into account the charge distribution given in Eq. (1) and the incident electric field $\boldsymbol{E}(r) = E_o\hat{z}$ of the microwave radiation propagating along the x-axis and polarized parallel to the z-axis (see **Figure 2**), considering the size of the particle to be extremely small compared to microwave radiation wavelength, is obtained to be:

$$\boldsymbol{f} = \rho_1(\boldsymbol{r})E_o\hat{z} \tag{4}$$

Operating on the Eq. (2) the $\nabla.$ operator from the left-hand side, substituting Eq. (3), Eq. (4) and rearranging the terms the following wave equation is obtained:

$$\nabla^2 P_1 + k_1^2 P_1 = E_o u_o z \tag{5}$$

where

$$k_1 = \frac{\omega/c_1}{\sqrt{1+j\varepsilon}}, = \frac{\omega\chi}{c_1^2\rho_1}, u_o = -\frac{10Q}{3\sigma\alpha^2} \tag{6}$$

The pressure field outside of the particle assuming an ideal fluid is described by two respective equations of Eqs. (2) and (3):

$$j\omega\rho_o\boldsymbol{v}_o(\boldsymbol{r}) = -\nabla P_o(\boldsymbol{r}) \tag{7}$$

$$j\omega P_o(\boldsymbol{r}) = -c_o^2\rho_o\nabla.\boldsymbol{v}_o(\boldsymbol{r}) \tag{8}$$

where $P_o(\boldsymbol{r})$ is the pressure, $\boldsymbol{v}_o(\boldsymbol{r})$ the velocity, $\rho_o$the mass density and c_o the acoustic speed. Also combining Eqs. (7) and (8):

$$\nabla^2 P_o(\boldsymbol{r}) + k_o^2 P_o(\boldsymbol{r}) = 0 \tag{9}$$

where $k_o = \omega/c_o$ is the wave constant of the infinite space outside of the sphere.

5. Solution of the boundary value problem

5.1 The field expression inside the sphere r<α

The acoustic pressure inside the spherical particle being excited by the interaction of microwave electric field component acting to electric charges, being inside the spherical volume, could be described in terms of the primary ($P_{10}(\boldsymbol{r})$) and secondary ($P_{11}(\boldsymbol{r})$) pressure fields [33].

In the analysis to follow spherical coordinates are used r, θ and φ being the radial distance from the origin, θ being the angle measured from z-axis and φ the azimuth angle.

The primary field $P_{10}(\boldsymbol{r})$ should satisfy Eq. (4) with the right-hand side inhomogeneous term. Based on Green's theory, assuming the outside medium being infinite, the primary pressure is determined by using the equation:

$$P_{10}(\boldsymbol{r}) = u_o \iiint_{Sphere} G_1(\boldsymbol{r}, \boldsymbol{r}')\, z' d\boldsymbol{r}' \tag{10}$$

inserting $z'=r'cos(\theta')$ and the expansion [34].

$$G_1(\boldsymbol{r}, \boldsymbol{r}') = -jk_1 \sum_{n=0}^{\infty} j_n(k_1 r_<) h_n^{(2)}(k_1 r_>) \sum_{m=-n}^{n} Y_n^m(\theta, \varphi) Y_n^{*m}(\theta', \Phi') \tag{11}$$

where $j_n(.)$ and $h_n^{(2)}(.)$are the spherical Bessel and Hankel (second type) functions,$r_< = \min(r, r')$ and $r_> = \max(r, r')$, the angular spherical wave function

$$Y_n^m(\theta, \varphi) = j^n \sqrt{\frac{2n+1}{4\pi} \frac{(n-m)!}{(n+m)!}} e^{jm\varphi} P_n^m(\theta, \varphi) \tag{12}$$

and $P_n^m(\theta, \varphi)$being the Legendre function.

Substituting Eq. (11) into Eq. (10) and $z'=r'cos(\theta')$, the fact the double summation in Eq. (10) being limited on the terms m=0 and n=1 after the orthogonality of the angular wave functions and the Bessel functions integral [35]

$$\int_{r=0}^{\alpha} j_1(k_1 r_<) h_1^{(2)}(k_1 r_>) r'^3 dr' = a^4 j_1(k_1 r) w_o + jr/k_1^3$$

$$\text{whith } w_o = \left(3h_1^{(2)}(k_1 a) - k_1 a h_0^{(2)}(k_1 a)\right) / \left(a^2 k_1^2\right)$$

leads to the result of the primary pressure field

$$P_{10}(\boldsymbol{r}) = jk_1 u_o E_o \cos(\theta) \left(a^4 j_1(k_1 r) w_o + \frac{jr}{k_1^3} \right) \tag{13}$$

Noticing that the primary field depends only to $P_1^o(cos\theta) = \cos(\theta)$ angular function (n=1 and m=0 terms), the secondary pressure field is written easily:

$$P_{11}(\boldsymbol{r}) = A\, j_1(k_1 r) \cos(\theta) \tag{14}$$

5.2 The field expression outside the sphere (r>α)

Considering the excitation of only the wave with $\cos(\theta)$ dependence and the necessity of radiation condition to be valid for $r \to +\infty$ we can write easily:

$$P_o(\mathrm{r}) = B\, h_1^{(2)}(k_1 r) \cos(\theta) \tag{15}$$

In Eqs. (14), (15), the unknown coefficients A and B are determined by imposing the validity of the boundary conditions at the spherical surface $r=\alpha$:

Continuity of pressure fields $P_{10} + P_{11} = P_0$

Continuity of radial velocities $\hat{\boldsymbol{r}}.(\boldsymbol{v_1}(\boldsymbol{r}) - \boldsymbol{v_0}(\boldsymbol{r})) = 0$

Then the A and B coefficients are calculated easily after some algebraic operations.

The final solution for the secondary field inside the sphere and in particular for the total pressure is obtained to be:

$$P_1 = a^{-2}QE_o\cos(\theta)W \tag{16}$$

$$W = -j2.569k_1a\left[\left(j_1(k_1r)w_o + \frac{jr}{a^4k_1^3}\right) + j_1(k_1r)S\right]$$

$$S = T/R$$

$$T = \left(ak_1j_1'(k_1a)w_o + \frac{j}{a^3k_1^3}\right)h_1^{(2)}(k_oa)a^{-1}k_o^{-1} - (j_1(k_1a)w_o + ja^{-3}k_1^{-3})h_1'^{(2)}(k_oa)$$

$$R = j_1(k_1a)h_1'^{(2)}(k_oa) - (1 + j\varepsilon)j_1'(k_1a)h_1^{(2)}(k_oa)\frac{\rho_ok_1}{\rho_1k_o}$$

6. Numerical calculations

After computing the pressure field as given in Eq. (16), it is shown that the "form factor" W is a function of the dimensionless quantities k_1a,k_oa,$(\rho_ok_1)/(\rho_1k_0)$, r/α and the parameter related to the viscosity of the virus particle $\varepsilon = \frac{\omega\chi}{c_1^2\rho_1} = \left(\frac{k_oac_o^2}{c_1^2}\right)\delta$, where $\delta = \chi/(c_o\rho_1\alpha)$ is a quantity related to the total viscosity of the spherical virus particle. Remembering that $\chi = \frac{4\eta}{3} + \kappa$ [36], where η,κ are the shear and bulk viscosity coefficients of a Newtonian fluid, we take the quantity δ as a "measure of the degree of viscosity" in our calculations. Furthermore, the interest being on the maximum pressure on the particle, we take $\theta = 0$ or π on the two poles of the sphere where the rupture of the virus capsid is sought.

In **Figure 3**, numerical results of the W "form factor" are given in the range $0.1 < k_oa < 3.0$ for various parameters of the ratios $\frac{\rho_o}{\rho_1} = 1.05,1.1,1.2$ and 1.3, c_0=1560m/s (speed of sound in the outer space), c_1 = 1950 m/s (speed of sound inside the sphere) . Meanwhile, the viscosity parameter is taken χ = 0.01 and χ = 0.001, which corresponds to the total viscosity constant corresponding to δ = 0.1 and δ = 0.01 ($N{\bullet}s/m^2$). The numerical results show interesting resonance behavior when viscosity is δ = 0.01. The phenomenon is stronger as δ decreases.

It is well known that the scattering of incident waves to a sphere (acoustic or electromagnetic waves) shows resonance phenomena when the refractive index of the spherical scatter has a large value. This phenomenon is well known in classical and quantum physics (Regge poles). As mentioned in the introduction section several researchers have foreseen this phenomenon. However, in the present analysis, the adopted model and analysis takes into account, although in a simplified form, all the involved mechanisms. The resonance is occurring near the angular frequency $\omega = \pi c_o/(2a)$, which corresponds to the "dipole mode" of the spherical particle.

In order to assess the feasibility of utilizing the phenomenon to compete with the virus populations, we need to calculate the pressure being developed at the spherical surface. Placing in Eq. (16) $r = \alpha$, 2α = 100 nm and $Q = Ne_o$, e_o=1.62 x 10^{-19} Cb (electron charge), N being the number of + or – charges we obtain after Eq. (16) the

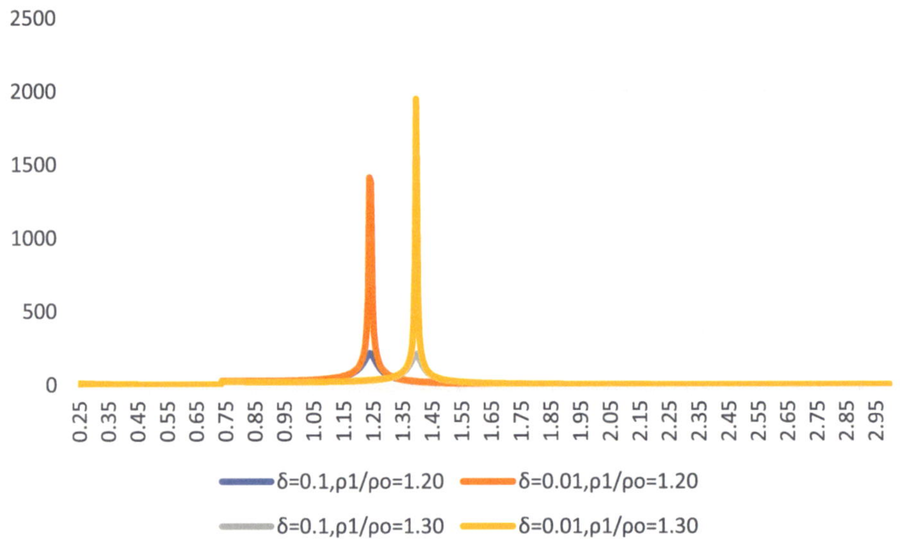

Figure 3.
Dependence of W (Eq. (16)) function to $k_o\alpha$. *The W function is the "form factor" of the resonating spherical virus particle and the resonance takes place when the R term takes maximum value at a specific microwave frequency, which is inducing mechanical-acoustic oscillation through electric force interaction between electromagnetic wave and electric charges.*

pressure $P_1 = 4 \bullet 10^{-5} \bullet N\, E_o\, W$, since $W \sim 2.000$ (see **Figure 3**) following the data given in ref. [9], the surface electric charge being $\sigma \leq 0.5eo/nm^2$ the virus area being $A_s = 4\pi\alpha^2$, we obtain $N \sim 3 \bullet 10^4$ and $P_1 \sim 2.400\, E_o$ (Pa). Then, if the imposed electric field at microwave frequency is E_o = 1.000 (V/m) (this corresponds to a power density of 130 mW/cm^2 much less used in hyperthermia treatments many times being 15.000 mW/cm^2 as a continuous wave signal), we arrive to the estimation that the pressure oscillation amplitude exerted on the two poles of the virus will be $P_1 \sim 2.4$ MPa. The

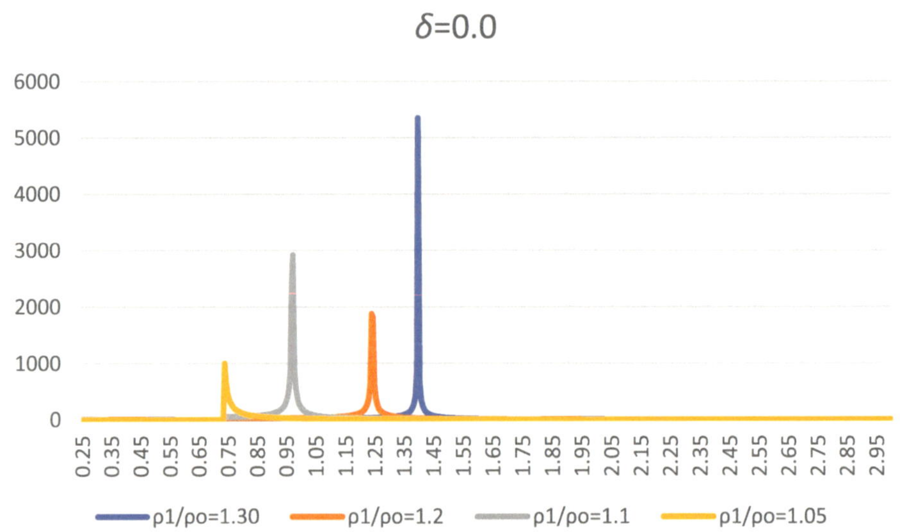

Figure 4.
Dependence of W [Eq. (16)] function to $k_o\alpha$ *in case of absence of viscosity.*

mentioned microwave field-generated pressure wave on the capsid surface seems to be comparable with the bulk Young's module being equal to 5 MPa (see the end of the section 'Methods' of ref. [8]).

In **Figure 4**, computations in the case of zero viscosity are presented for various ratios ρ_1/ρ_o. The strong resonance phenomenon of the ratio ρ_1/ρ_o is 1.3 while the lowering of the resonance frequency as this ratio decreases while the peak value of W has some variation.

The above initial results show that the argument expressed by the National Taiwan University in their seminal paper of ref. [18] is verified theoretically with the present model.

The microwave resonance frequency is computed easily known the value of $k^*= k_o a$ the peak value is attained, that is $f_{resonance}\ (Hz)=k^*c/(2\pi\alpha)$ where $c=3.10^8$ (m/s) is the speed of electromagnetic waves in vacuum.

7. Conclusions

A simplified model of a virus particle allowed to analyze the coupling phenomena between microwave (electromagnetic) radiation and acoustic waves generated inside the particle. Based on recent publications on virus physical and electronic properties of viruses, similar to Covid-19, computations show the possibility of strong interactions to generate rupture or capsid of the viruses. This action is based on the Coulomb force exerted by the oscillating field on the inhomogeneous electric charges within the spherical particle. The microwave resonance frequency—which is identical to the acoustic wave—is in the region of 6–10 GHz.

The prospect of using the presented principle to sterilize public spaces, hospitals, clinics etc. is an attractive proposition. The present microwave technology is available for the development of this type of portable device. Moreover, the existing more than 40 years of experience in clinical hyperthermia, which is based on the use of low microwave frequencies as an adjuvant therapy to treat many cancer diseases, makes it attractive to investigate the possibility of developing technologies to implement the mentioned idea in the future to depopulate virus populations inside the human body. Before this, extensive in vitro trials in virus and cell cultures need to be carried out to follow with animal trials as well.

Author details

Nikolaos K. Uzunoglu
National Technical University of Athens, Athens, Greece

*Address all correspondence to: nuzu@central.ntua.gr

References

[1] Mateu MG. Mechanical properties of viruses analyzed by atomic force microscopy: A virological perspective. Virus Research. 2012;**168**(1–2):1-22

[2] Pedro J. Atomic force microscopy of virus shells. DOI: 10.1016/j.semcdb.2017.08.039

[3] Cieplak M, Robbins MO. Nanoindentation of 35 virus capsids in a molecular model: Relating mechanical properties to structure. DOI: 10.1371/journal.pone.0063640

[4] Guerra P, Valbuena A, Querol-Audí J, Silva C, Castellanos M, Rodríguez-Huete A, et al. Structural basis for biologically relevant mechanical stiffening of a virus capsid by cavity-creating or space filling mutations. Scientific Reports. 2017;**7**:4101

[5] Xian Y, Karki CB. The roles of electrostatic interactions in capsid assembly mechanisms of giant viruses. International Journal of Molecular Science. 2019;**20**:4101

[6] Juan F, Vicente-Alique E, Núñez-Ramírez R, Wang Y, Martínez-Salazar J. Evidences of changes in surface electrostatic charge distribution during stabilization of HPV16 virus-like particles. PLOS ONE. 2016;**371**:490

[7] Rees F, Comas-Garcia M, Koay MST, Cornelissen JJLM, Knobler CM, Gelbarta WM. Role of electrostatics in the assembly pathway of a single-stranded RNA virus. Journal of Virology. 2014;**88**:10479

[8] Šiber A, Podgornik R. Role of electrostatic interactions in the assembly of empty spherical viral capsids. Physical Review. 2007;**76**:0619

[9] Bozic AL, Siber A. Electrostatics-driven inflation of elastic icosahedral shells as a model for swelling of viruses. Biophysical Journal. 2018;**115**:822

[10] van der Schoot P, Bruinsma R. Electrostatics and the assembly of an RNA virus. Physical Review. 2005;**71**:0619

[11] Yang X, El Baroudi A, Le Pommellec JY. Analytical approach for predicting vibration characteristics of an embedded elastic sphere in complex fluid. Archive of Applied Mechanics

[12] Vladimir A. Fonoberov and Alexander A. Balandin, Low-frequency vibrational modes of viruses used for nanoelectronic self-assemblies. Physica Status Solidi. 2004;**241**:R67

[13] Elhafi G, Naylor CJ, Savage CE, Jones RC. Microwave or autoclave treatments destroy the infectivity of infectious bronchitis virus and avian pneumovirus but allow detection by reverse transcriptase-polymerase chain reaction. Avian Pathology. 2004;**33**:303

[14] Wu Y, Yao M. In situ airborne virus inactivation by microwave irradiation. China Science Bulletin. 2014;**59**:1438

[15] Bristow I, Lim WC, Lee A, Holbrook D, Savelyeva N, Thomson P, et al. Microwave therapy for cutaneous human papilloma virus infection. EJD. 2017;**27**:5

[16] Siddharta A, Pfaender S, Malassa A, Doerrbecker J, Engelmann M, Nugraha B, et al. Inactivation of HCV and HIV by microwave: A novel approach for prevention of virus

transmission among people who inject drugs. Scientific Reports. 2017;**6**:366

[17] Babincová M, Sourivong P, Babinec P. Resonant absorption of ultrasound energy as a method of HIV destruction. Medical Hypotheses. 2000; **55**(5):450-451

[18] Yang S-C, Lin H-C, Liu T-M, Lu J-T, Hung W-T, Huang Y-R, et al. Efficient structure resonance energy transfer from microwaves to confined acoustic vibrations in viruses. Scientific Reports. 2015;**5**:1038

[19] Tamura A, Ichinokawa T. Frequency spectrum of a small particle. Journal of Physics. 1983;**16**:83

[20] Ahadi A, Johansson D, Evilevitch A. Modeling and simulation of the mechanical response from nanoindentation test of DNA-filled viral capsids. Journal of Biological Physics. 2013;**39**:183-199

[21] Liu T-M, Chen H-P, Yeh S-C, Wu C-Y, Wang C-H, Luo T-N, et al. Effects of hydration levels on the bandwidth of microwave resonant absorption induced by confined acoustic vibrations. Applied Physics Letters. 2009;**95**:173702

[22] Burkhartsmeyer J, Wang Y, Wong KS, Gordon R, Trapping O. Sizing, and probing acoustic modes of a small virus. Applied Physics Letters. 2020;**10**:394

[23] Eric C, Sankey OF. Atomistic modeling of the low-frequency mechanical modes and Raman spectra of icosahedral virus capsids. Physical Review E. 2010;**81**:2191

[24] Mousavizadeh L, Ghasemi S. Genotype and phenotype of COVID-19: Their roles in pathogenesis. Journal of Microbiology, Immunology and Infection

[25] Hung W-T, Tung J-J, Chen S-Y. A focusing reflectarray and its application in microwave virus sanitizer. Research Article. 2013;**21**:481

[26] Kouloulias V, Plataniotis G, Kouvaris J, Dardoufas C, Gennatas C, Uzunoglu N. Chemoradiotherapy combined with intracavitary hyperthermia for anal cancer feasibility and long-term results from a Phase II Randomized Trial. American Journal of Clinical Oncology. 2005;**28**:91

[27] Ono R. Bronchoscopic Local Nd:YAG Laser Hyperthermia in the Treatment of Lung Cancer. Diagnostic and Therapeutic Endoscopy. 1995;**1**:159-164

[28] Cottis PG, Uzunoglu NK. Focusing properties of dipole arrays placed near a multilayer Lossy sphere. Journal of Electromagnetic Waves and Applications. 1990;**4**:431-440

[29] Nikita KS, Mitsis GD, Uzunoglu NK. Analysis of focusing of pulsed baseband signals inside a layered tissue medium. IEEE Transactions on Microwave Theory and Techniques. 2000;**48**(1):30-39

[30] Lamb H. On the Vibrations of an Elastic Sphere. Proceedings of the London Mathematical Society. 1881;**13**: 189

[31] Available from: https://www.who.int/health-topics/coronavirus#tab=tab_1

[32] Davis JL. Wave Propagation in Solids and Fluids

[33] Sommerfeld A. Partial Differential Equations in Physics. New York: Academic Press; 1949

[34] Philip M. Morse and Herman Feshbach, Methods of Theoretical Physics. New York: McGraw Hill; 1958

[35] Abramowitz M, Stegun IA. Handbook of Mathematical Functions. London: Dover Publ; 1972

[36] Ghosh T, Prasad D, Dutt N, Rani KY. Viscosity of Liquids: Theory, Estimation, Experiment, and Data. Springer

Chapter 6

Assessment of the Addition of Fluorapatite in Hydroxyapatite Coatings: Implementation Prosthetics/Bone in Vivo

Halima Feki Ghorbel, Awatef Guidara, Yoan Danlos, Jamel Bouaziz and Christian Coddet

Abstract

Hydroxyapatite (Hap: Ca_{10} $(PO_4)_6$ OH_2)-Fluorapatite (Fap: Ca_{10} $(PO_4)_6F_2$) composite coating on 316 L stainless steel, using the High-Velocity Oxy-Fuel Spray (SHVOF), was investigated. This work is an evaluation of the bioactivity of bone/Hap–Fap composite coatings implanted in the tibia of the rabbit. A small amount of Fap (6.68, 13.26 and 26.52 w% Fap attributed to 0.25, 0.5 and 1% fluor) was introduced into Hap. The fluorapatite provides more stable and adherent deposits. The characteristics of the coatings were investigated with various instruments including Scanning Electron Microscopy (SEM) and X-Ray Diffraction (XRD) and biological (in-vivo and in-vitro) tests. Hap–Fap coating showed excellent behavior in vitro and in vivo tests revealing that the Fap is effective in improving biocompatibility and bioactivity. This study draws inspiration from technological and biological selection solutions adopted by evolution, to transpose the principles and processes of human engineering.

Keywords: Fluorapatite, hydroxyapatite, high-velocity oxy-fuel spray, In-vitro/In-vivo test

1. Introduction

Research has never ceased trying to improve materials for bone and dental implants and the techniques used to synthesize them. To begin with, metals were the most widely adopted materials for implants. However, they were not without problems. Indeed, Schulze et al. assessed the effect of In-Vivo exposure to metallic nanoparticles on bone marrow In-Vitro, revealing significant alterations in cell biology [1]. Similarly, in Total hip arthroplasty 'THA', the problem of implant loosening due to aseptic osteolysis, was observed to be triggered by wear particles from the implant articulating surfaces. The overall In-Vivo performance of metallic hip implants turned out to be comparably poor and caused some manufacturers to recall their products. Thus, several researchers turned to the study of various bioactive ceramics, such as Hydroxyapatite (Hap), Tricalcium phosphate (TCP) and Bioglass, developed to activate bone regeneration [2]. Given its similar chemical and physical characteristics to bone, Hap was widely used as bone substitute material in orthopedic domain [3, 4]. However, when dispersed as nanoparticles Hap can cause inflammation by activating monocytes and neutrophils [4–6].

Within this prospect, Fluorapatite (Fap) was initially chosen for this purpose because of its fairly good biological and mechanical properties compared to β-TCP and hydroxyapatite (Hap) when used alone. Nevertheless, the long-term stability of bioactive ceramic implants was criticized for at least two shortcomings: the presence of low solubility of the coating and the absence of high adhesion strength between coating and substrate [7]. Hence, as a solution to these problems, Dhert and Cheng proposed to maintain F-concentration down to the minimum level in order to decrease the solubility of fluorohydroxyapatite (FHa) [8, 9]. In a similar attempt to decrease the solubility of the coating, Wolke et al. recommended the use of a fluoride coating containing Hap [10]. Indeed, Hap and Fap constitute the inorganic compound of the human hard tissue. Unfortunately, Baltag et al. and Overgaard et al. reported that the high degradation rate of Hap coating in biological environments is a serious concern, which might be harmful to adhesion properties, resulting in undesirable debris and even delaminating, which eventually leads to the failure of the implant [11, 12]. Furthermore, the stability of Hap–Fap composites was also a source of concern because dissolution or reprecipitation influences cell behavior [13]. Indeed, when Hap–Fap composites are suspended in SBF liquid, they can dissolve and precipitate rapidly, leading to ion release, change in the pH, size, and morphology [14, 15].

As a solution to these problems, some studies [16, 17], presented Fap as potential replacement for Hap in implants because of the higher degradation of the latter material in biological environments and its lower adherence. For these reasons, other researchers worked on developing the mechanical properties of coatings from Fap and Hap together to increase their efficiency. Bahandag et al. [17] studied the influence of Fap on the properties of Hap coating using thermal projection. These scholars revealed that Fap increases Hap coating crystallinity. Furthermore, the slow release of Fap reduces the delivering particles leading to the improvement of its osteointegration. Similarly, Chang et al. [18] showed that the fluorine ions improve the osteoblastic cells' proliferation and differentiation. Hence, scholars such as Fraz and Telle and Somrani opted for Fap as an additive to Hap and a replacement for pure Hap coating on metallic implants because of their chemical composition which is similar to the bone mineral [19, 20]. Mark and Brown confirmed this finding and explained that these materials presented good biocompatibility [21]. Particularly in dental prostheses, fluor is effective in inhibiting caries [22] and proves to be compatible with the human bone that contains approximately 1 wt% of fluor [23–25].

The second issue related to implants was the technique of synthesizing the chosen materials. Thermally sprayed bio-ceramic on metallic substrate was widely used in orthopedic prostheses given its great potential in bone regeneration activity In-Vivo. Among the adopted Ceramic coating techniques are plasma spraying, flame spraying and high-velocity oxy-fuel (HVOF) spray [26]. The Plasma spray process is the most commercially preferred technique for clinical applications. The use of bond coats to improve the surface properties of metallic substrates was also studied extensively in the thermal spray literature [27]. However, despite the abundance of research on Hap coating and on the mechanical properties and bioactivity of Fap, there is still a serious need to improve the synthesis, characterization and application of Fap–Hap coating composite. For this reason, this work intended to investigate the Fap (6.68; 13.26 and

26.52 w%)-Hap coating on the 316 L stainless steel using the SHVOF technique. The remainder of this paper is divided into three sections. Section two will present the characterization of raw materials and of as-sprayed apatite coating. Section three will exhibit the study of the bioactive behavior of samples using the simulated body fluid (SBF). Finally, section four will discuss the adhesion and biocompatibility of Hap and Hap-Fap coating on rabbit bone cells as applied through the SHVOF technique.

2. Materials and methods

2.1 Materials

The Fap powder was synthesized using a wet-chemical method [20]. A calcium nitrate solution was slowly poured into a boiling solution containing di-ammonium hydrogen-phosphate, and a 28% NH_4OH solution was added to the mixture in order to adjust the pH to 9. The precipitate was filtered, washed, dried at 70°C for 12 h and calcinated at 500°C. The Hap powder with Ca/P ratio of 1.66–1.71 (Medicoat, HA-15-183, 95%) was used as base material. The Hap–Fap powders were mixed using a dried-mechanical method.

2.2 Methods

The Hap and Hap–Fap coatings were carried out on 316 L stainless steel discs by High-velocity oxy-fuel spray (SHVOF) using a Sulzer-Metco F4-MB (Switzerland). Before applying the coating, degreasing and grit blasting were carried out to make the substrate surface coarse and clean. The thermal sprayed process parameters are listed in **Table 1**. The coating thickness was evaluated using a micrometer (200 μm).

The simulated body fluid (SBF) which has ionic concentrations very similar to those of human plasma was used to study the bioactive behavior of samples. For this, a commonly used SBF solution of pH 7.4 was prepared according to the procedure recently described by Kokubo et al. [28]. **Table 2** presents the ionic composition of the as-prepared SBF and compares it with that of human plasma. The Hap and Hap–Fap coatings were cut in parallelepipedic slices of size 2*6*12 mm and then cleaned before being immersed in 100 ml of SBF. The temperature was maintained at 37°C.

The phase compositions of the coated samples were examined with an X-ray diffractometer (PHILIPS PANALYTICAL) and with a scanning electron microscope (JEOL JSM 5800LV). The roughness of the surfaces of both substrate and coatings was measured using Mitutoyo (ISO 1997).

Prosthesis implantation, for an "In Vivo" study in rabbits, was carried out in a laboratory of animal experimentation and approved by the faculty of medicine of Sfax and by the ethical committee of the Habib-Bourguiba University Hospital, Sfax;

Starting powders	Ch_4flow rate (L/min)	N_2 flow rate (L/min)	O_2 flow rate (L/min)	Spray flow rate (ml/min)	Spray distance (mm)
Hap	135	20	270	100	130
Hap-Fap	135	20	270	100	130

Table 1.
Spray process parameters.

Ion concentration (mmol/l)	SBF	Blood plasma
Na^+	142	142
K^+	5.0	5.0
Mg^{2+}	1.5	1.5
Ca^{2+}	2.5	2.5
Cl^-	147.8	103.0
HCO^-	4.2	4.2
$HPO^{2-}{}_4$	1.0	1.0
$SO^{2-}{}_4$	0.5	0.5
pH	7.25–7.42	7.24–7.40

Table 2.
Ion concentrations in supersaturated SBF solution prepared in the present study and in human blood plasma.

Tunisia. The surgical operation was performed on the rabbit, by an appropriate medical team with respect for all the rules of asepsis. The shape and dimensions of prostheses are shown in **Figure 1**. They were carefully chosen after several graft tests in order to perfectly fill in the medullar canal of the Rabbit tibia.

These prostheses were then sterilized by ^{60}CO gamma irradiation (Equinox, UK).

Five adult white rabbits, aged from 5 to 10 months, were used for the experiments. All surgical procedures were done under strict aseptic protocol. The animals were premedicated and anesthetized with a xylasine/ketamine mixture (10 mg/kg). The animal tibialis anterior face was shaved. After the injection, the animal was left at rest with its eyes closed for about 20 minutes. The skin was disinfected with a povidone-iodine solution of 10% (Betadine, Medapharma). Local anesthesia (Unicaine 2%) was employed in the anterior tibialis face. Four cylindrical bars with a length of 37 mm and diameter of 4–5 mm (**Figure 1**) were placed in each animal. Surgical preparations for the cylinders were done using first a pilot drill and then a 2 mm twist drill. Careful drilling was done with a low rotary handpiece. The skin was sutured with interrupted

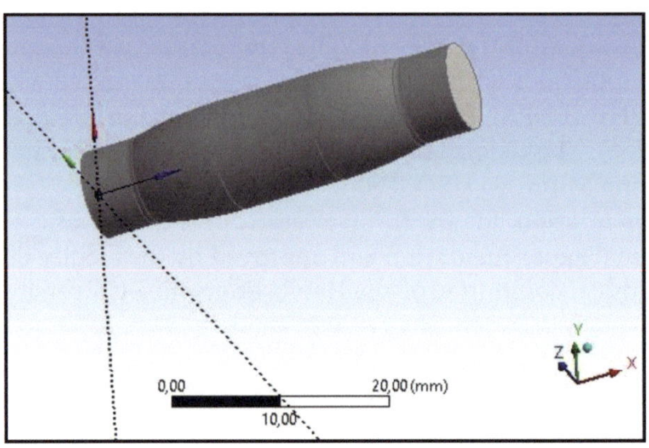

Figure 1.
Prostheses dimensions for implantation: Great diameter: 5 mm; small diameter: 4 mm; degree of inclination: 30° and length: 37 mm.

threaded sutures. Skin was again soaked with a povidone solution. After 28 days the rabbits were sacrificed, and the implants extracted. The samples were stored in a formalin-based solution called BB'S and then included in methacrylic resin to be used for radiological testing of bone tissue. X-ray radiography was performed using a Faxitron X-ray system (Edimex, Angers, France) equipped with a camera (5X5 CCD). The coatings were characterized by Scanning Electron Microscopy (SEM) and X-Ray Diffraction (XRD). Moreover, the bonding strength of the as-sprayed coatings was measured using a universal testing system. The microstructure of the detached surfaces was examined using SEM.

3. Results and discussion

3.1 Characterization of the different powders

Figure 2 shows the XRD patterns for Hap and Hap-synthesized Fap composite. This observation can have four implications. Firstly, Hap-diffraction peaks shown in **Figure 2a** can be indexed as typical hexagonal phases (ICDD 77–0120). These peaks situated at approximately 25.9, 32, 33.1, 40.1 and 46.7 correspond well to (002), (211), (300), (212) and (222) lattice planes of the classic hexagonal phase Hap-diffraction peak (a) can be indexed as typical hexagonal phase (ICDD 77–0120). Secondly, it was clearly observed that the addition of Fap in Hap did not affect the diffraction peaks location shown in **Figure 2b–d**. However, the increase of Fap quantity in Hap powders shown in **Figure 2b–d** slightly increased the diffraction peaks intensity. This observation, in total agreement with the standard requirements, indicates that the crystallinity was well preserved. Thirdly, this observation confirms Feki-Ghorbel et al.'s previous claim that the crystalline stability of Fap is greater than that of Hap [29]. Finally, this observation shows that there were no impurity diffraction peaks or phases in the XRD patterns of the Hap and Hap–Fap.

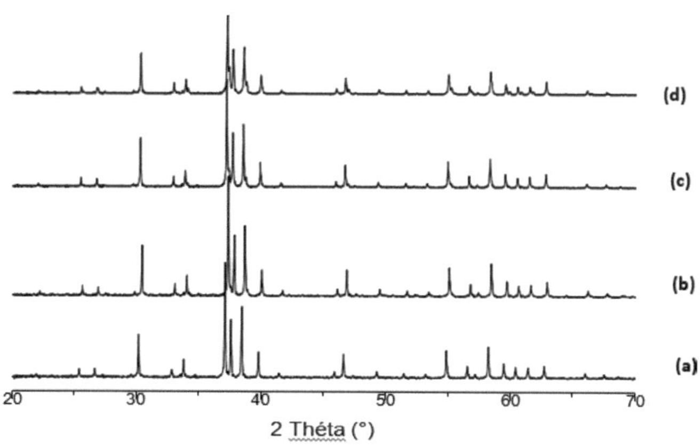

Figure 2.
XRD patterns of powders: (a) hap; (b) hap–Fap (0.25 w% F); (c) hap–Fap (0.5 w% F) and (d) hap–Fap (1 w% F).

3.2 Characterization of as-sprayed apatite coating

3.2.1 SEM analysis

The SEM micrograph shown in **Figure 3** reveals a characteristic lamellar microstructure of thermal spray coatings. A typical surface microstructure HVOF sprayed Hap coatings are generally porous. The porous structure might be beneficial to the

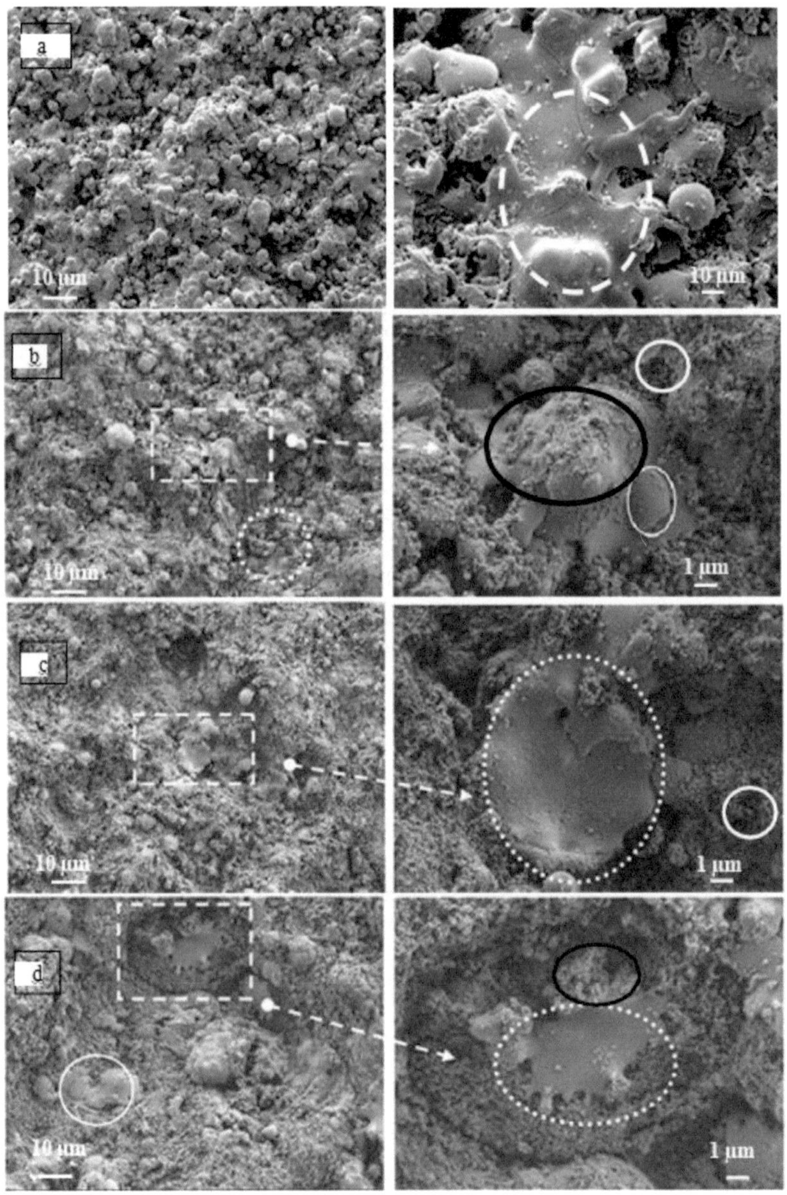

Figure 3.
Surface microstructures of hap (a); hap–Fap (0.25% F) (b); hap–Fap (0.5% F) (c) and hap–Fap (1% F) (d).

biomedical application involving the mechanical fixation by bone in growth. Firstly, indicate where you can see the microstructures in terms of **Figure 3a–c**.

Secondly, these observations were not the result of the same quantity of Fap. Therefore, the differences in microstructures can be the result of the quantity of Fap. This should be clearly explained. The microstructure of the Hap and Hap–Fap coatings consists of particles of different shapes and sizes: fully molten splats which are merged in each other (white circles in the photos), some small globular particles in the matrix give a dense appearance and others probably are the un-melted particles. No crack can be observed.

3.2.2 XRD analysis

Figure 4 shows the XRD patterns for Hap and Hap–Fap as-sprayed coating. We can observe the presence of the pattern peaks attributed to Fap and Hap. Diffraction peaks of Hap–Fap coating showed more intensity than that of Hap coating. This can be interpreted as an indication that the addition of Fap increases the crystalline phase of Hap coating. Hence, our XRD analysis confirms once more Feki-Ghorbel et al.'s previous claim that the crystalline stability of Fap is greater than that of Hap [29]. In addition, this result suggests that the HVOF spray parameters were optimized. Finally, the observation of a little amorphous phase implies that Fap can be chosen as a potential partial substitute for Hap.

3.2.3 Surface roughness

The surface properties of implants are of vital importance for implant tissue interaction which further influences the biocompatibility for clinical use [30, 31]. Particularly, surface roughness alters osteoblastic attachment proliferation of bone cells and their differentiation and matrix production [32, 33]. The surface roughness parameters (Ra, Rq and Rz) for blasted stainless steels and for 316 L substrates coated by Hap and Hap–Fap are shown in **Figure 5**. The HVOF spray treatment clearly modified the surface roughness of the samples (Ra). In fact, the average surface roughness of blasted steel substrate increased from1.5 to 2.9 ± 0.2 μm. Moreover, the addition of Fap improved the

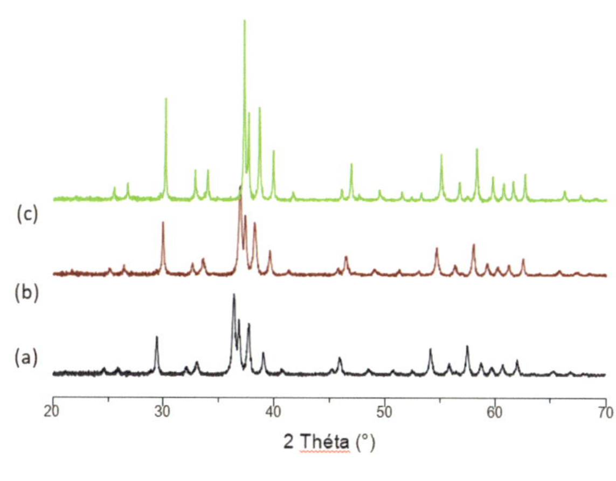

Figure 4.
XRD patterns of powders: (a) hap; (b) hap–Fap (0.25 w% F), (c) hap–Fap (1 w%F).

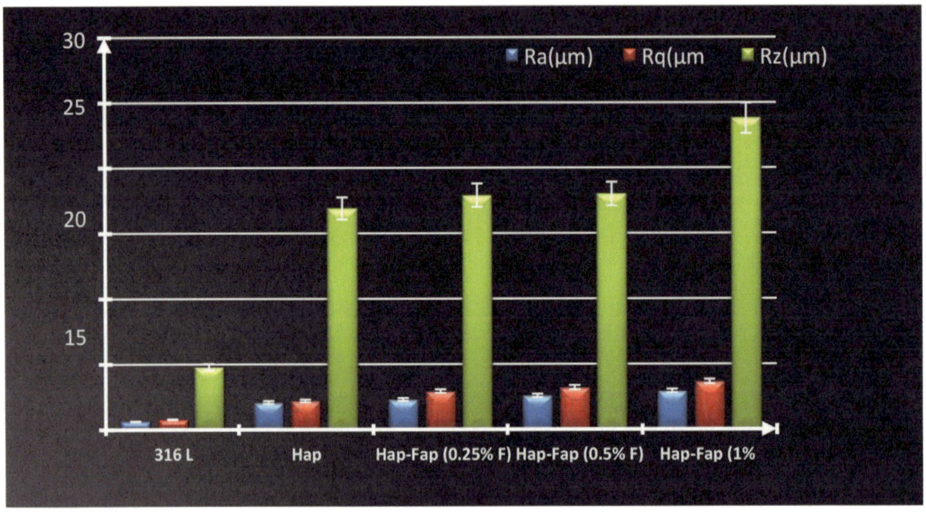

Figure 5.
Surface roughness measurements for the SHVOF sprayed hap; hap-Fap-coated 316 L along with that of the grit blasted uncoated 316.

average surface roughness which reached 4.7 ± 0.2 µm for the Hap–Fap (w% 1 fluor). Such results are relevant for the clinical application of Hap–Fap coated implants. As was reported by authors [34, 35], the modification of the surface roughness of an implant significantly influences In-Vitro osteoblastic response. Furthermore, a better long-term In-Vivo response of the implant is achieved when the surface roughness is increased as the amount of bone in direct contact with the implant surface as well as the loads and torques required for extracting the implant from bone growth increase [28].

The comparison between the surface roughness results obtained in this work and those of the reported HVOF-sprayed and the flame-sprayed Hap coating [36–38] shows that the Hap–Fap coatings presented a comparative surface roughness while the addition of Fap yielded a higher roughness of the surface.

3.3 In-vitro tests

3.3.1 XRD analysis

Figure 6 presents the XRD patterns of SHVOF Hap–Fap (0.25 and 1 w% F) composite coatings in the function of immersion time in SBF. After each immersion time in SBF (0, 3 and 28 days), the examination of the coating surface shows the expansion of the diffraction lines group around 37° (2θ) corresponding to Hap phase. After a 28-day immersion in SBF, the XRD pattern presented a diffraction halo situated between 35° and 40° (2θ). The mineralogical phase analysis, using the software "High score", allowed identifying the transition phases appearing on the Hap–Fap surface. The diffraction patterns can be attributed to a hydrated carbonated apatite layer. These patterns included lines of diffraction which can be mainly attributed to a carbonated apatite [00–012-0529]. An intensity peak around 34.5° which can be attributed to a.

Hydrated Apatite [01–077-0128] was also detected. The crystalline structure modification after a 28-day immersion in SBF produced a formation of a new carbonated apatite precipitate covering all the Hap–Fap (0.25 w% and 1 w% F) surface.

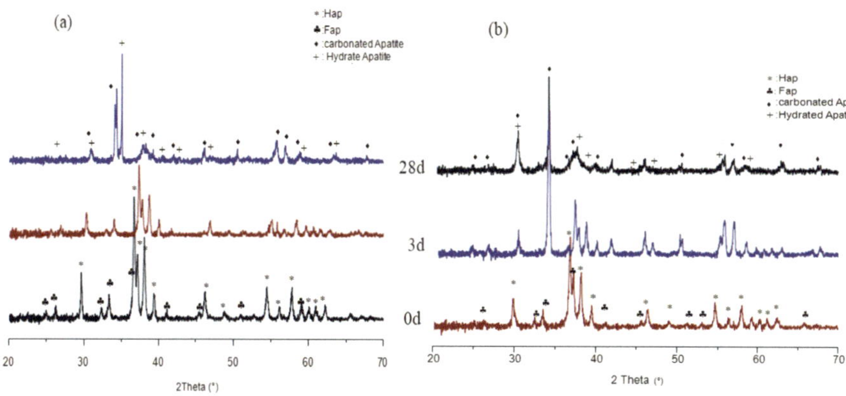

Figure 6.
XRD patterns of SHVOF hap–Fap composite coating in function immersion time in SBF: 0.25 w% F (a) and 1w% F (b).

3.3.2 SEM morphologies

The bioactivity and biocompatibility of the Hap coating and Hap–Fap composite coatings were evaluated by immersing the samples in SBF for 0; 3; 7 and 8 days.

As shown in **Figure 7**, spherical-shaped particles were observed on the Hap surface of the coated material after three days of immersion in SBF. **Figures 8** and **9** confirmed with certainty the appearance of the new crystals on the Fap phase of the Hap–Fap composite on Hap–Fap at 0.25 and 1w% composite. As the immersion time increased, the crystals seemed to grow in size and to form an Apatite layer as shown after A 28-day immersion in SBF. This layer appeared to consist of many nano-sized crystallites with spherical morphology, as was observed by previous authors [39]. Feki-Ghorbel et al. [29] revealed that the incorporation of 1 w% of Fap into Al_2O_3 promoted the apatite layer formation on the coating surface when soaked in SBF. According to Kokubo [28], the bone-bonding ability of a material depends on the ability of apatite to form on its surface in SBF. In light of this idea, we can say with much certainty that since we observed the appearance of the apatite crystals on the Fap phase of the Hap–Fap composite surface, then we expect to reach very encouraging results in the in vivo tests.

Figure 10 presents the micrographs of the SHVOF coating/bone interface following sacrifice of the animal 28 days after the implantation.

The Hap (**Figure 10a**) and Hap–Fap (**Figure 10b,c**) composite coating exhibits a good adhesion with bone. The addition of Fap encourages the presence of crystalline forms interconnected and constituting an array at the Hap–Fap/Bone interface.

A Cross-section observation also confirms the presence of a thin apatite layer on the Hap-composite surface as shown by In-vitro tests. In this work, we have observed that the incorporation of a little amount of Fap (1w%) into Hap promoted the apatite layer formation on the coating surface and could achieve bone-bonding when implanted in bone tissues, as do bioactive ceramics.

The understanding of the biological mechanisms involved in osteoconduction seems complex. In fact, the correlation between the microstructure and the biological activity

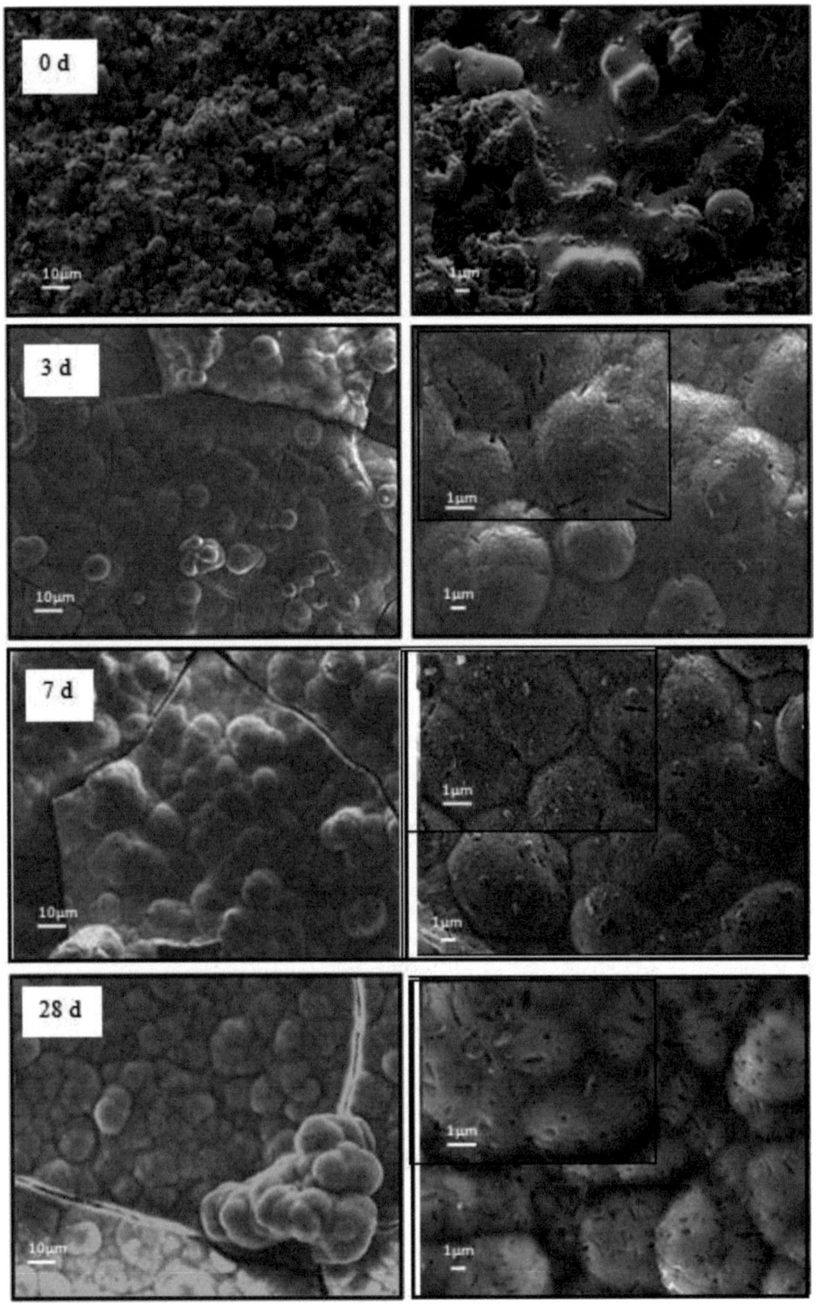

Figure 7.
SEM observation of hap surface after immersion in SBF solution.

can intervene to be able to interpret the results. Nevertheless, two facts should be kept in mind when dealing with calcium phosphate films. Firstly, as was previously stated [40, 41], the bioactivity of the calcium phosphates coating depends on the capacity of their surface nucleated crystallized carbonated apatite, like the osseous mineral from the biological fluids. Secondly, the calcium phosphate films are crystallized under

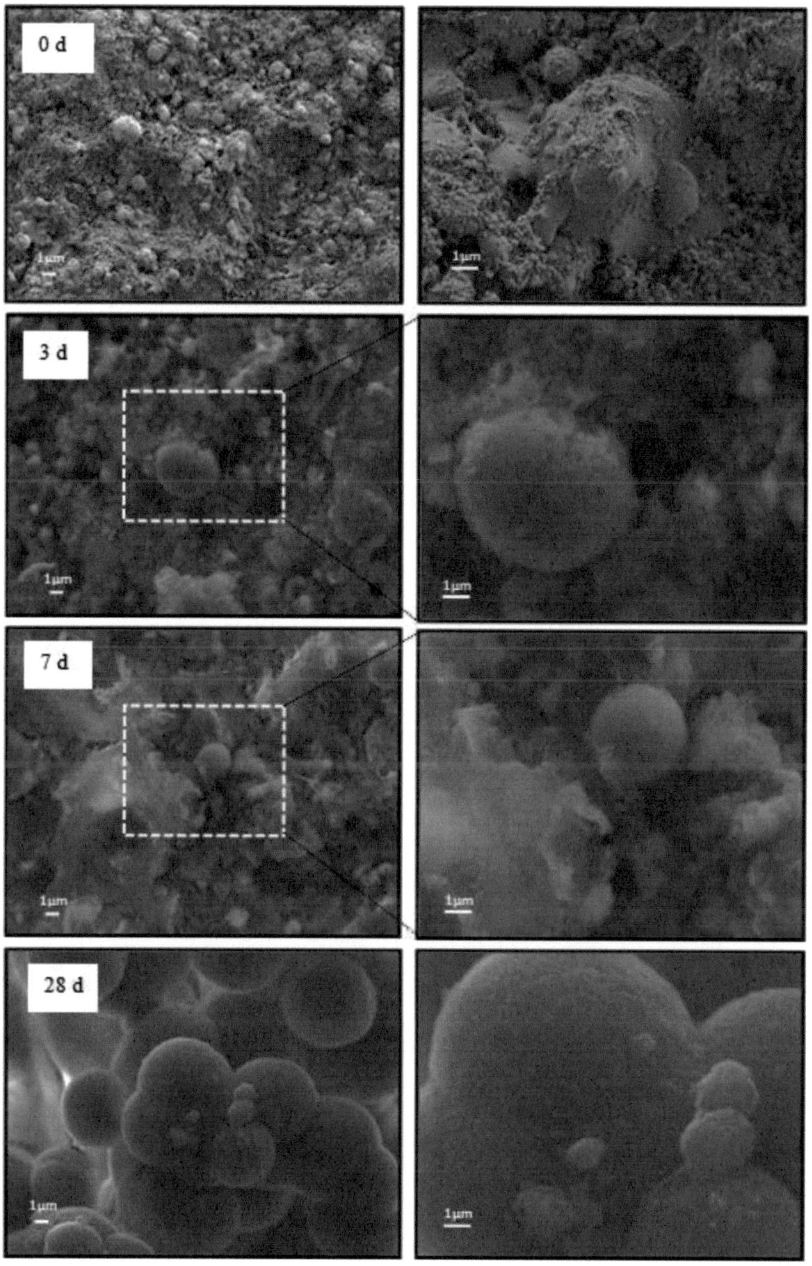

Figure 8.
SEM observation of hap–Fap (0.25 w% F) surface after immersion in SBF solution.

conditions such as the formation of the random nucleated spherulites, which can be considered as only a stage essential to the mechanism of the osteo-conduction. Hence, it can be concluded that the structural composition of the biomaterial's surface has an influence on bioactivity. In this study, the addition of the various percentages of Fap in the Hap coating acted on the surface's physical state and so increased its bioactivity with nucleated apatite spherulites on the biomaterial surface. As regards Hap/Fap composite

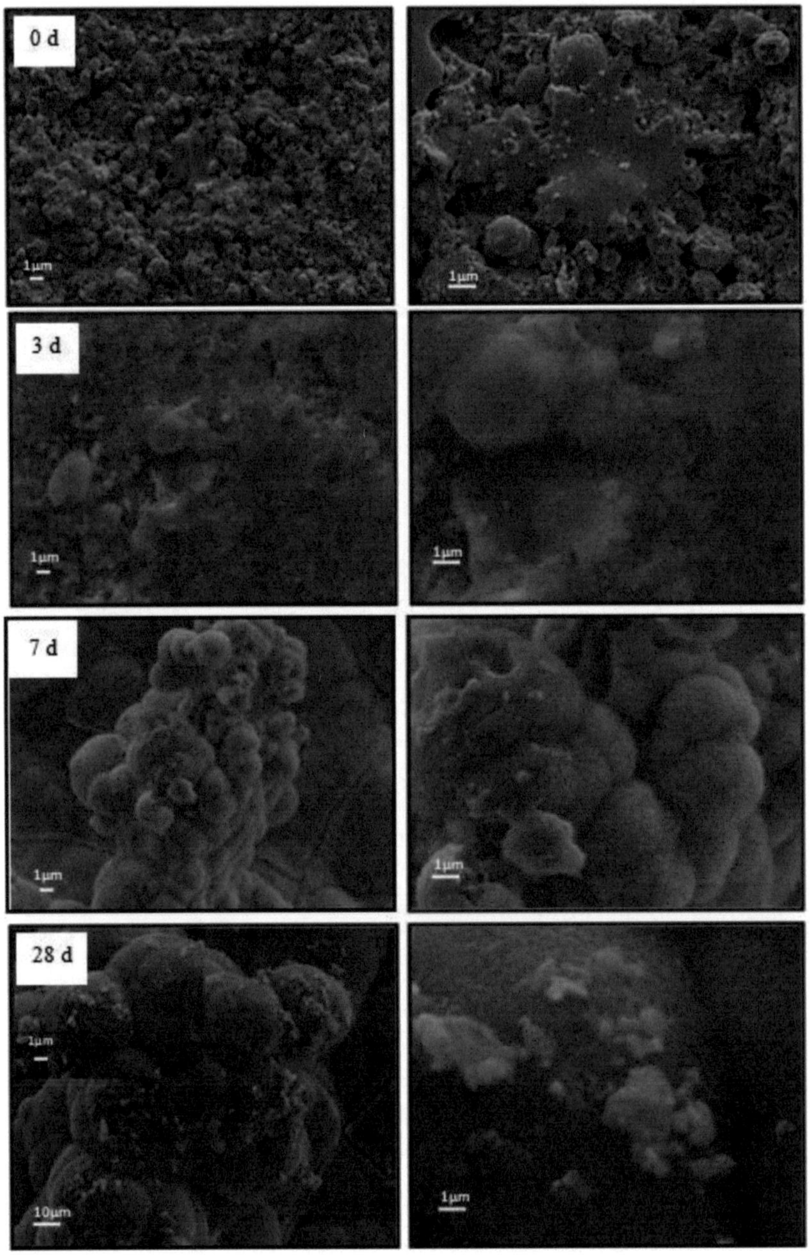

Figure 9.
SEM observation of hap–Fap (1w% F) surface after immersion in SBF solution.

coatings, after 28 days of immersion in the SBF and in contents raised in Fap (1% F), the observed transition phase can prove the presence of an amorphous neo-formed structure on the surface justifying the samples bioactivity.

Indeed, the important solubility of the amorphous phase increases the SBF oversaturation when in contact with the sample surface. Therefore, it triggers the nucleation of a neo-formed film similar to the osseous mineral. The SEM study of the

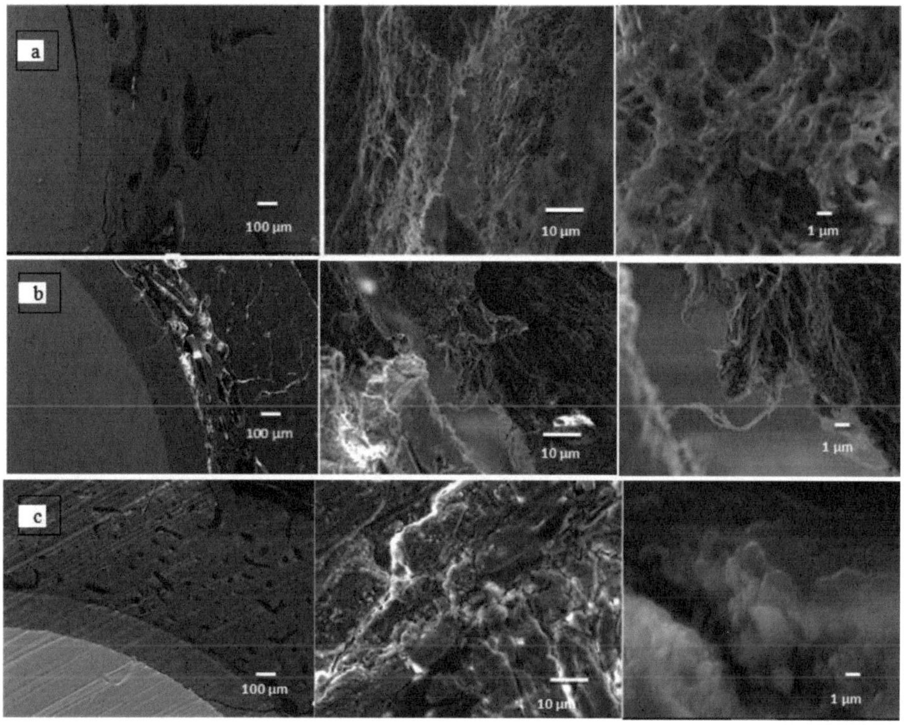

Figure 10.
SEM observation of a. hap/bone interface (a); hap–Fap (0.25w%F)/bone interface (b) and hap-Fap (1 w% F) / bone interface (c).

surface film morphology after 3, 7 and 28 days of incubation showed a large amount of extra-cellular matrix proliferation.

This matrix stems from the crystallized apatite layer which triggered the osseous regrowth. It is important to note that this new apatite phase will result in osteoconduction. The physicochemical reactions occurring on the surface of bioactive materials could be decomposed into several stages described in the in vitro analysis above. Thus, the development of a layer of amorphous calcium phosphate occurred after a stage consisting of relegated ions forming a network of modifiers present in the deposits of the Hap–Fap matrix.

In chemical terms, this implies that Ca^{2+} ions present on the bone surface are quickly exchanged with the ions H+ present in the SBF. Hence, the amorphous precipitates of phosphate of calcium grow on the surface. In addition, the migration of the ions Ca^{2+} on the bone surface, the presence of PO_4^{3-} in the Hap–Fap matrix and the existence of soluble phosphates in the SBF promote the bioactive phase formation.

At this stage, the amorphous calcium phosphate layer covers the coating surface containing an important amount of bioactive Fap. The calcium phosphates are crystallized thanks to the presence of the ions OH-, Mg^{2+} and carbonates CO_3^{2-} in the middle. Then, the apatite carbonated phase occurs close to the bone mineral phase.

A Previous study, [16, 17] presented Fap as potential replacement for Hap in implants because of the higher degradation of the latter material in biological environments and its lower adherence.

4. Conclusions

To conclude, this investigation resulted in four important findings.

Firstly, this study demonstrated that the addition of Fap would produce a higher roughness at the bone surface which would certainly enable the osteoblastic cells to grow more easily. This would be very likely due to the fact that the more the surface of the bone is rough, the more it provides direct contact with the implant surface and the more it requires loads and torques for extracting the implant.

Secondly, the SEM observation of In Vitro tests confirms the formation of an Apatite layer on Hap and Hap–Fap coating after a 28-day immersion in SBF. The addition of 0.5w% and 1 w% Fap in the Hap coating increases the bone surface bioactivity through the growth of nucleated apatite spherulites on the biomaterial surface. In addition, it reduces the Hap phase solubility and leads to a stable amorphous layer.

Thirdly, this study revealed that the exchange of the calcium ions " Ca^{2+} " and the phosphate ions " $PO4^{3-}$ " between the Hap-Fap matrix and the SBF promotes the bioactive apatite phase formation. This new apatite phase leads to osteo-conduction.

Fourthly, In Vivo tests showed partial resorption of the implant in the form of zones less dense than the bone. The presence of this area would indicate the new bone growth accompanied by a slight osteointegration.

Our results suggest that the addition of Fap into Hap is more suitable for implantation of prostheses and for the study of bone substitutes. We can conclude with certainty that it represents the solution of choice the principles and processes of human engineering.

The bioactive behavior comparison of the two composites Hap-Fap with 0.25 wt% F and 1 wt% F revealed that the increased Fap level until 1 wt% F is presented as a better choice because it allowed the implant to develop a better integration which will guarantee a more durable life.

Acknowledgements

We would like to thank the medical team at the University hospital of Habib Bourguiba, Sfax, Tunisia for their valuable help in conducting the in vivo implantation on the rabbit in their laboratory.

Author details

Halima Feki Ghorbel[1,2]*, Awatef Guidara[1], Yoan Danlos[2], Jamel Bouaziz[1] and Christian Coddet[2]

1 LCI, Ecole Nationale d'Ingénieurs de Sfax"ENIS", Tunisia

2 LERMPS, Université de Technologie de Belfort-Montbéliard, France

*Address all correspondence to: ghorbel.halima@gmail.com; halima.ghorbel-feki@uvsq.fr

References

[1] Schulze F, Gramoun A, Crowe LA, Dienelt A, Akcan T, Hofmann H. Accumulationofamino-polyvinylalcohol-coated superparamagnetic ironoxide nanoparticles in bone marrow: Implications for local stromal cells. Nanomedicine. 2015;**10**:2139-2151

[2] Chu TMG, Orton DG, Hollister SJ, Feinberg SE, Halloran JW. Mechanical and in vivo performance of hydroxyapatite implants with controlled architectures. Biomaterials. 2002;**23**(5):1283-1293

[3] de Groot K. Bioceramic consisting of calcium phosphate salts. Biomaterials. 1980;**1**:47-50

[4] Palmer SP, Chen S, Rubino S, Weng H, Xia W, Engqvist H, et al. In vivo and in vitro evaluation of hydroxyapatite nanoparticle morphology on the acute inflammatory response. Biomaterials. 2016;**90**:1-11

[5] Velard F, Laurent-Maquin D, Guillaume C, Bouthors S, Jallot E, Nedelec JN, et al. Polymorphonuclear neutrophil response to hydroxyapatite particle, implication in acute inflammatory reaction. Acta Biomaterialia. 2009;**5**:1708-1715

[6] Laquerriere AG, Laquerriere P, Laurent-Maquin D, Guenounou M, Phillips TM. The effect of the physical charactertics of hydroxyapatite particles on human myocytes IL-18 production in vitro. Biomaterials. 2004;**25**:5921-5927

[7] Ding SJ, Ju CP, Lin JHC. Journal of Biomedical Materials Research. 1999;**47**:551

[8] Dhert WJA, Thomson P, Klein CPTA, de Groot K, Rozing PM, Ericson LE. Journal of Materials Science: Materials in Medicine. 1994;**5**:59-66

[9] Cheng K, Weng W, Han G, Du P, Shen G, Yang J, et al. Materials Chemistry and Physics. 2003;**78**:767-771

[10] Wolke JGC, Dhert WJA, Klein CPTA, de Groot K, de Blieck-Hogervorst JMA. Characterization of plasma-sprayed fluoroapatite coatings for biomedical applications. Materials Science Monographs. 1991;**69**:285-294

[11] Baltag I, Watanabe K, Kusakari H, Taguchi N, Miyakawa O, Kobayashi M, et al. Long-term change of hydroxyapatite-coated dental implants. Journal of Biomedical Materials Research. 2000;**53**:76-85

[12] Overgaard S, Lind M, Josephsen K, et al. Resorption of hydroxyapatite and fluorapatite ceramic coatings on weight-bearing implants: A quantitative and morphological study in dogs. Journal of Biomedical Materials Research: An Official Journal of The Society for Biomaterials, The Japanese Society for Biomaterials, and the Australian Society for Biomaterials. 1998;**39**(1):141-152

[13] Wang L, Nancollas GH, Henneman Z, Klein E, Weiner S. Nanosized particles in bone and dissolution insensitivity of bone mineral. Biointerphases. 2006;**1**:106-111

[14] Tang LWR, Nancollas GH. Size-effects in the dissolution of hydroxyapatite an understanding of biological demineralization. Journal of Materials Chemistry. 2004;**14**:2341-2346

[15] Thomann JM, Voegel JC, Gramain P. Kinetic of dissolution of calcium hydroxyapatite powder. III:

pH and sample conditioning effects. Calcified Tissue International. 1990;**46**(10):121-129

[16] Zhang S, Wang YS, Zeng XT, Khor KA, Weng W, Sun DE. Thin Solid Film. 2008;**516**:5151-5748

[17] Bhadang KA, Gross KA. Influence of Fluorapatite on the Properties of Thermally Sprayed Hydroxyapatite Coatings. Vol. 69. Australia: School of Physics and Materials Engineering, Monash University; 2004. p. VIC 3800

[18] Chang YL, Lew D, Park JB, Keller JC. Biomechanical and morphometric analysis of hydroxyapatite-coated implants with varying crystallinity. Journal of Oral and Maxillofacial Surgery. 1999;**57**:1096-1108

[19] Somrani S. Mise en forme d'apatites nanocristallines: céramiques et ciments [thesis]. University of Tunisia; 1990

[20] Franz ED, Telle R. Reaction hot pressing of Fluorapatite for dental implants. In: Vincenzini P, editor. High Tech Ceram. Germany. 1987. pp. 31-41

[21] Mark F, Brown PW. Phase relationships in the ternary system CaO—P2O5—H2O at 25°C. Journal of the American Ceramic Society. 1992;**75**(12):3401

[22] Osborn JF, Neweslety H. Biomaterials. 1980;**1**:908

[23] Ben Ayed F, Bouaziz J. Sintering of Tricalcium phosphate-Fluorapatite composites by addition of alumina. Ceramics International. 2008;**34**:1885-1892

[24] Agathopoulos S, Tulyaganov DU, Marques PAAP, Ferro MC, Fernandes MHV, Correia RN. The fluorapatite– Anorthite system in biomedicine. Biomaterials. 2003;**24**:1317-1331

[25] Gross KA, Rodrıguez-Lorenzo LM. Biodegradable composite scaffolds with an interconnected spherical network for bone tissue engineering. Biomaterials. 2004;**25**:48-49 4955-4962

[26] Sobieszczyk S. Surface modifications of Ti and its alloys. Advances in Materials Science. 2010;**10**(1):29-42

[27] Heimann RB. Plasma Spray Coating Principles and Applications. New York: VCH Weinheim; 1996

[28] Kokubo T, Kushitani H, Sakka S, Kitsugi T, Yamamuro T. Solutions able to reproduce in vivo surface-structure changes in bioactive glass-ceramic A-W. Journal of Biomedical Materials Research. 2002;**24**:721-734

[29] Feki-Ghorbel H. Elaboration et caractérisation des couches minces de composite Apatite-Alumine sur un support métallique par projection thermique [thesis]. France, Tunisia: UTBM University, Sfax University; 2016

[30] Schwartz Z, Boyan BD. Underlying mechanisms at the bone-biomaterial Interface. Journal of Cellular Biochemistry. 1994;**3**:340-347

[31] Lee TM, Tsai RS, Chang E, Yang CY, Yang MR. The cell attachment and morphology of neonatal rat Calvarial osteoblasts on the surface of Ti6Al4V and plasma sprayed HA coating: Effect of surface roughness and serum contents. Journal of Materials Science. Materials in Medicine. 2002;**13**:341-350

[32] Kieswetter K, Schwartz Z, Hummert TW, Cochran DL, Simpson J, Dean DD, et al. Surface roughness modulates the local production of growth factors and

cytokines by osteoblast-like MG-63 cells. Journal of Biomedical Materials Research. 1996;**32**(1):55-63

[33] Lincks J, Boyan BD, Blanchard CR, Lohmann CH, Liu Y, Cochran DL, et al. Response of MG63 osteoblast-like cells to titanium and titanium alloy is dependent on surface roughness and composition. Biomaterials. 1998;**19**:2219-2232

[34] Aparicio C, Engel E, Gil FJ, Planel JA. Journal of Materials Science: Materials in Medicine. 2002;**13**:1105

[35] Aparicio C, Gil FJ, Fonseca C, Barbosa M, Planell JA. Biomaterials. 2003;**24**:263

[36] Schmid R. Comportement des matériaux dans les milieux biologiques. Application en médecine et biotechnologues. Première édition ed. Lausame in Swiss; 1999

[37] Gross KA, Babovic M. Influence of abrasion on the surface characteristics of thermally sprayed hydroxyapatite coatings. Biomaterials. 2002;**23**(24):4731-4473

[38] Fernnandez J, Gaona M, Guilemany JM. Effect of heat treatments on HVOF hydroxyapatite coatings. Journal of Thermal Spray Technology. 2007;**16**(2):220-228

[39] Feki Ghorbel H, Guidara A, Danlos Y, Bouaziz J, Coddet C. Synthesis and characterization of alumina-fluorapatite coatings deposited by atmospheric plasma spraying. Materials Letters. 2017;**185**:268-271

[40] Rey C, Renugo V, Palakrishmanand B, Glimcher MJ. Fourier transform infrared spectroscopic study of the carbonate ions in bone mineral during aging. Calcified Tissue International. 1991;**49**:251-258

[41] Al Kattan A, Errassifi F, Sautereau AM, Sarda S, Dufour P, Barroug A, et al. Medical potentialities of biomimetic Apatites through adsorption, ionic substitution, and mineral/organic associations: Three illustrative examples. Advanced Engineering Materials. 2010;**12**(7):1438-1656

Section 4

Biomimetic Mechanicobiology

Chapter 7

Perspective Chapter: Viscoelastic Mechanical Equivalent Models

Emad Kamil Hussein, Batool Mardan Faisal, Kussay Ahmed Subhi, Thiago Santos, Samir Ghouali, Muhammad Asyraf and Caroliny Santos

Abstract

Today, we are living in a polymeric era where thousands of daily used products are manufactured from some polymeric materials with different tasks and under a wide range of ambient conditions, including time duration of loading and working condition temperature. This leads to focusing light spot on behavior of such specific materials and investigating the strain associated with the applied stress to understand both of creep and stress relaxation behavior of the loaded polymeric components. Hence, this chapter deals with the estimation of induced strain allied with the applied force on a polymeric material via establishing the so-called mechanical equivalent models starting from the simple elastic element (spring with a modulus of elasticity E), simple viscous element (damper or dashpot with fluid viscosity η), Maxwell model, Voigt model, modified Maxwell model, modified Voigt model, and Maxwell-Voigt model. The theoretical analysis was built on derivation of the prompted deformation, as a function of time in each of the employed models, as a result of the applied external load (force) and then by depending on Hook's law transforming the gained expressions into stress (σ) and strain (ε) notation, followed by comparing the obtained equation with the general formula of the Hook's law to find exact values of the constant and as coefficients of the stress and strain. Final theoretical analysis showed that Maxwell's modified model was the best describing behavior of a loaded polymeric material to some extent followed by the other models.

Keywords: polymers, mechanical equivalent models, Maxwell model, Voigt model, creep, stress relaxation

1. Introduction

Rheology is a branch of physical sciences concerned in the wide sense with the deformation and flow of materials. Whereas, theoretical rheology aims to establish the general laws of rising and development in time of deformation and investigate the general properties of processes on a strictly mathematical basis. Applied rheology establishes a bridge between the theoretical results and practical applications by introducing certain additional simplifying assumptions. The rheological properties of real materials are determined qualitatively and quantitatively by experimental rheology, which supplies the theory with new ideas and constitutes an ultimate basis of its verification [1]. Observation of the physical facts and their superficial description, without looking for deeper causes of observed phenomena, is essential to formulate a phenomenal approach in rheology. However, the final of rheology, as a science, is to establish the relation of physical causes given conditions of deformation and flow, to the known properties of the constituent particles of materials aspects of rheology based on the nature of inter-atomic forces and the structure of the matter. The concerns of rheology, with problems of flow and deformation of materials, its ranges of interest are conventional, thus any kind of physical effect, which form their definition is instantaneous (time-dependent) are considered from a rheological point of view as particular or limiting cases. Especially, this is concerned with elastic and plastic types of deformation. The classical theory of elasticity is founded on a linear dependence between stress and strain and its time independence. Thus, when a loaded elastic body exhibits an instantaneous response to the applied stress and if the physical causes are removed, then the strain is fully recoverable. The assumption about the smaller elastic strain allows us to apply the superposition principle for both mechanical variables, the stress and the strain [2]. However, in investigating the mechanical behavior of different conditions, behave themselves in accordance with the assumptions stated for elastic body. For example, when applying constant stress for an extended time intervals, the resulting strain increases in time. On the other hand, constant strain find a time-dependent decrease in stress. Thus, it is found that mechanical properties of certain groups of materials are variable with time [3]. Moreover, it states that there is no one-to-one correspondence between stress and strain as for an elastic body and at an arbitrary time-instant. The mechanical variables depend on the past history of straining and stressing, respectively.

2. Creep and stress relaxation

In particular, the viscoelastic materials have the ability to increase their deformation in time by constant stress is called creep process, and the property of stress drop-in time by a constant strain is the stress relaxation process, see **Figure 1.**

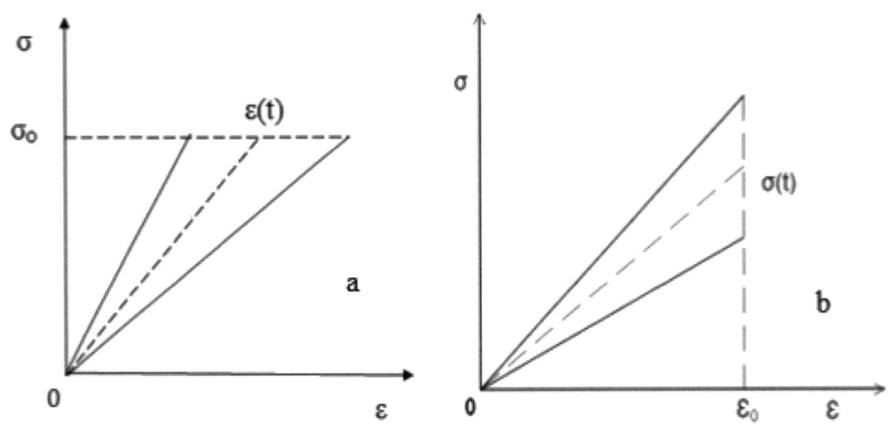

Figure 1.
(a) Creep and (b) stress relaxation mechanical behavior.

Both of the above definitions are used in a narrower sense rather to specify a two-time dependent physical function that describes the characteristic features of visco-elastic behavior relaxation and creep. In general, both phenomena occur at any variable stress and strain even simultaneously. In describing the rheological phenomena by means of mathematical formulation, we usually follow some general principles which are in accordance with our physical experience, on the other hand for particular materials we make some constitutive assumptions on an experimental basis [4]. What are restrictive conditions on the possible rheological processes? The physical relations obtained in such a way are so-called constitutive equations or equations of state in more or less general way stress and strain as the physical causes and the physical effort respectively. The aim of deriving the constitutive equations is to characterize and classify as adequately as possible the real material properties known from the experimental data. The constitutive equation must be in general in agreement with the two fundamental principles which secure its invariance. The first one is the principle of objectively material properties. It simply states that the material properties are objective and cannot be dependent on the observer and lies point of view, no matter how lies position is. Thus, according to the principle, if it is found that certain rheological processes are described by a constitutive equation, then every process equivalent to it is compatible with the same constitutive equation two equivalence of processes are stated on the basis of the transformation relation of space and time [5].

In order to determine the relation of stress to a rheological process, we state in agreement with physical experience the local character of stress. Thus, the stress at a material particle depends on what happens only in an arbitrary small vicinity of the particle. The distant parts of the body do not have a direct influence on the value of stress at the particle considered. Further, we can use causality principle. It expresses the fact that any physical process, at an arbitrary time-instant, may depend on what occurred in all past instants, that is, on the past history of happenings only. These two principles give rise to the concept of determinism for stress. According to this principle, the stress at a material particle in an arbitrary time is determined by the past history of the rheological process in an arbitrary small vicinity of the particle. In certain cases, we also assume some restrictive conditions on the possible motion during a rheological process. These conditions are constitutive restraints connected with the general features of geometry of the possible motion, for example the assumption of in compressibility of a media.

The condition of incompressibility implies that every possible motion is isochoric, that is, the deformational motion of rheology of a body occurs with a constancy of volume. Thus, the density of the body does not change during the process considered. Constitutive equations are presented in different mathematical formulation for instance in the form of differential and integral equations and as functional.

In general, the differential form of constitutive equation contains strain, strain rate, stress, stress rate, and higher derivatives of both strain stress with respect to time. It is also containing some explicit functions of time [3]. In the last case, the physical properties of the material are variable with time independently of existing stress state and must be given in advance. If there are some temperature changes, the influence of which should be taken into account in the constitutive equation may appear explicitly, temperature as a new variable [6]. Thus, in general, the differential form of a constitutive equation contains may be written in form

$$f(\varepsilon, \dot{\varepsilon}, \ddot{\varepsilon}, \dots \dots, \sigma, \dot{\sigma}, \ddot{\sigma}, \dots \dots \ \mathrm{t(time)}, T(\mathrm{temp.})) = 0 \qquad (1)$$

If the temperature T does not appear in the equation, the rheological process is said to be isothermal; on the other hand, if the time-variable t disappears explicitly in the equation, then the differential equation with constant coefficients.

3. Polymers

Polymers exist in nature in such forms as wood, rubber, jute, hemp, cotton, silk, wool, hair, horn, and flesh. In addition, there are countless man-made polymeric products, such as synthetic fibers, engineering plastics, and artificial rubber. In certain aspects, the deformation of polymeric solids bears a strong resemblance to that of metals and ceramics. Polymers become increasingly deformable with increasing temperature, as witnessed by the onset of additional flow mechanisms [7]. Also, the extent of polymer deformation is found to vary with time, temperature, tress, and microstructure, constituting parallel deformations for fully crystalline solids. Furthermore, time–temperature equivalence for polymeric deformation is indicated, which is strongly reminiscent of the time–temperature parametric relations that will be discussed the basic features of the polymeric structure that dominate flow and fracture properties will be also discussed.

4. Viscoelastic response of polymers

The deformation process of many materials depends to a varying degree on both time-dependent and time-independent processes. It is known that when the test temperature is sufficiently high, a test bar would creep with time under a given load. Likewise, were the same bar to have been stretched to a certain length and then held firmly, the necessary stress to maintain the stretch would gradually relax such response is said to be viscoelastic. Since the glass and melting temperatures and most of the polymeric materials are not much above ambient (and in fact may be lower as in the case of natural rubber) these materials exhibit viscoelastic creep and stress relaxation phenomena at room temperature (25°C). When the elastic strain and viscous flow rate are small (approximately 1 up to 2% and 0.1, respectively) the viscoelastic strain may be approximately by:

$$\varepsilon = \sigma f(t) \quad \text{Linear Viscoelasticity} \tag{2}$$

That is, (stress/strain) ratio is a function of time only. This response is called linear viscoelasticity and involves a simple addition of linear elastic and linear viscous (Newtonian) flow components [8].

When the stress–strain ratio of a material varies with time and stress, then:

$$\varepsilon = g(\sigma, t) \text{ Nonlinear Viscoelasticity} \tag{3}$$

Stress–strain ratio is a function of time only = Linear viscoelasticity.

Stress–strain ratio is a function of time and stress = Nonlinear viscoelasticity.

The viscoelastic response is nonlinear. A comparison of creep behavior between metals and polymers is clearly shown in **Table 1**.

On the basis of a simple creep test, it is possible to define a creep modulus as in the next equation:

$$E_c(t) = \frac{\sigma_o}{\varepsilon(t)} \text{ Creep modulus} \tag{4}$$

Creep behavior	Metals	Polymers
Linear elastic	No	Sometimes
Recoverable	No	Partially
Temperature range	High temperature above 0.2 T_h	All temperatures above 200°C

Table 1.
Metals and polymers creep behavior comparison.

where:
$E_c(t)$: Creep modulus as a function of time.
σ_o: Constant applied stress.
$\varepsilon(t)$: Time-dependent strain.
Now, a relaxation modulus $E_r(t)$ is defined as:

$$E_r(t) = \frac{\sigma(t)}{\varepsilon_o} \text{ Relaxation modulus} \tag{5}$$

where:
$E_r(t)$: Relaxation modulus as a function of time.
$\sigma(t)$: Time-dependent stress.
ε_o: Constant induced strain.
Both $E_c(t)$ and $E_r(t)$ moduli are varying with time as a time-dependent deformation; thus, the designer of a plastic component must look beyond the basis tensile test data when coupling the deformation response of a polymeric material. For example, for $\varepsilon_{\text{critical}}$ there is a linear relationship $\sigma = E\varepsilon_{\text{critical}}$ but this material will creep, that is, the level $\varepsilon_{\text{critical}}$ will increase with time, so to account for this additional deformation, the designer makes use of isochrounus stress–strain curves derived from creep data.

5. Mechanical models analogy

There is a strong similarity between behavior of some specific mechanical components, including (spring—elastic element and damper—viscous element) and the so-called viscoelastic materials, including polymerics, so it is better to define the basic two elements:
Spring as an elastic element, with stiffness K as indicated in **Figure 2**.
Damper is representing a viscous element with damping constant C exactly as illustrated in **Figure 3** below.

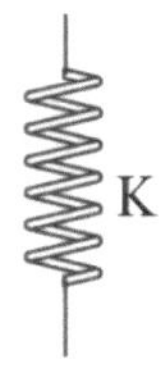

Figure 2.
Elastic element (linear spring) with stiffness constant K.

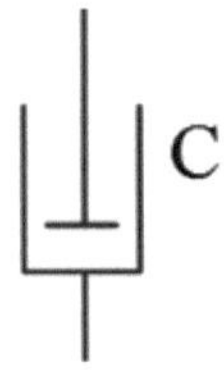

Figure 3.
Viscous element (damper) with damping coefficient C.

Based on alternative layout and assembly of the above two mentioned elements, there will be many equivalent models but the most popular equivalent mechanical models are listed below [9]:

1. Maxwell model
2. Voigt model
3. Modified Maxwell model
4. Modified Voigt model
5. Maxwell-Voigt model

6. Deformation process in polymeric solids

The following mathematical relationships are describing the mutual dependence of the main governing parameters, including the applied stress, the associated strain, and Young's modulus, in both cases of time-dependent and independent.

	Tensile test	Shear test	
Case (a)	$\varepsilon = \frac{\sigma}{E}$	$\gamma = \frac{\tau}{G}$	Time-independent behavior (6)
Case (b)	$\varepsilon = \frac{\sigma}{\eta}$	$\gamma = \frac{\tau}{\eta}$	Time-dependent behavior (7).

Where:
ε and γ: Tensile and shear strain rates.
σ and γ: Applied tensile and shear stresses.
η: Fluid viscosity in terms of stress-time.

It is better to use viscosity η instead of the damping coefficient C in the coming analysis. And the viscosity η is directly proportional to the ambient temperature T, ($\eta \propto T$) according to the Arrhenius-type relation as in the following equation:

$$\eta = \mathrm{A}e^{\Delta H \mathrm{RT}} \text{ Arrhenius equation} \tag{6}$$

Where:
ΔH: Viscous flow activation energy at a particular temperature.
T: Absolute temperature.
R: Universal gas constant.
A: Pre-exponential factor.

The viscosity depends on time, that is, at t = 0, the viscosity is extremely high, while at t goes to infinity, is small. In other words, deformation is purely viscous upon loading (t = 0) and is rigid to the dashpot consequently there is no strain associated with the same with time the viscous character of the dashpot element becomes evident as strain developed that is directly proportional to time. When the stress is removed, this strain remains. Now when the spring and dashpot are in series, as shown in **Figure 4** below, called Maxwell model, the mechanical response of the material possesses both elastic and viscous components [10], so the model is shown in the figure. Note that all the strains are recovered but the viscous strains arising from creep of the dashpot remain, since the elements are in series, the stress on each is the same and the total strain or strain rate is determined from the sum of the two components. Hence,

$$\frac{d\varepsilon}{\mathrm{dt}} = \frac{\sigma}{\eta} + \frac{1}{E}\frac{d\sigma}{\mathrm{dt}} \tag{7}$$

For stress relaxation conditions

$$\varepsilon = \varepsilon_o \text{ and } \frac{d\varepsilon}{\mathrm{dt}} = 0, \tag{8}$$

$$\frac{\sigma}{\eta} + \frac{1}{E}\frac{d\sigma}{\mathrm{dt}} = 0 \tag{9}$$

$$\frac{1}{E}\frac{d\sigma}{\mathrm{dt}} = -\frac{\sigma}{\eta} \tag{10}$$

and by using the separation of variables procedure:

$$\frac{d\sigma}{\sigma} = -\frac{E}{\eta}\mathrm{dt} \tag{11}$$

$$\int_{\sigma_o}^{\sigma} \frac{d\sigma}{\sigma}\frac{d\sigma}{\sigma} = -\int_0^t \frac{E}{\eta}\mathrm{dt} \tag{12}$$

$$\ln\sigma|_{\sigma_o}^{\sigma} = -\frac{Et}{\eta} \tag{13}$$

$$\ln\frac{\sigma}{\sigma_o} = -\frac{Et}{\eta} \tag{14}$$

Figure 4.
Maxwell mechanical model (spring and dashpot in series).

$$\sigma(t) = \sigma_o\, e^{-\frac{Et}{\eta}} \tag{15}$$

$$\sigma(t) = \sigma_o\, e^{-\frac{t}{t}} \tag{16}$$

$$\varepsilon_{\text{Total}} = \varepsilon_S = \varepsilon_D \tag{17}$$

$$\sigma_{\text{Total}} = \sigma_S + \sigma_D \tag{18}$$

$$\sigma_{\text{Total}}(t) = E\varepsilon + \eta \frac{d\varepsilon}{\text{dt}} \tag{19}$$

Where t is the relaxation time defined by $\frac{\eta}{E}$.

The extent of stress relaxation for a given material will depend on the relationship between and, so when >>, the material behaves elastically such that, in other words, when the spring and dashpot elements are combined in parallel as shown in **Figure 5**. (Voigt mechanical model), this unit predicts a different time-dependent deformation response. First, the strains in the two elements are equal and the total stress on the pair is given by the sum of the two components.

For creep test $\sigma_{\text{Total}}(t) = \sigma_o$ and after intefration, yields:

$$\varepsilon(t) = \frac{\sigma_o}{E}\left(1 - e^{-tt}\right) \tag{20}$$

The strain experienced by the Voigt model is shown in **Figure 6**. The absence of any instantaneous strain is predicted from this equation $\varepsilon(t) = \frac{\sigma_o}{E}\left(1 - e^{-tt}\right)$ and is related in physical sense to the infinite stiffness of the dashpot at t = 0. The creep strain seems to rise quickly thereafter but reaches a limiting value of $\frac{\sigma_o}{E}$ associated with the full extension of the spring under that stress. Upon unloading, the spring remains extended but now exerts negative stress on the dashpot. In other manner, the viscous strains are reversed, and in the limit when both spring and dashpot are unstressed, all the strains have been reversed. Consequently, the Voigt and the Maxwell models describe different types of viscoelastic responses [8].

A some more realistic description of polymer behavior is obtained with a four elements model consisting of Maxwell and Voigt models in series precisely as shown in **Figure 6**.

By combining the above three last equations, it can be seen that the total strain experienced by this model may be given by:

$$\varepsilon(t) = \frac{\sigma}{E_1} + \frac{\sigma}{E_2}\left(1 - e^{-tt}\right) + \frac{\sigma}{\eta}t, \tag{21}$$

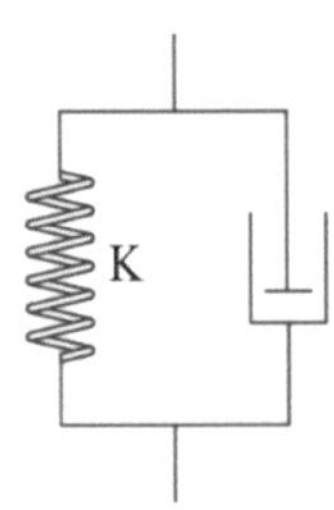

Figure 5.
Voigt mechanical model (spring and dashpot in parallel).

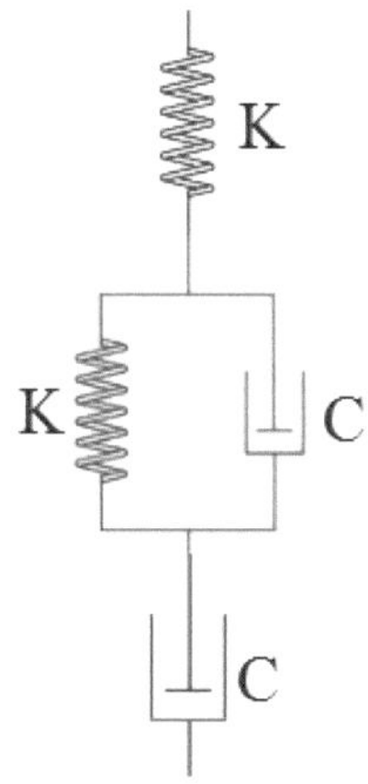

Figure 6.
Maxwell-Voigt mechanical model.

which takes account of elastic, viscoelastic, and viscous strain components, respectively.

The temperature dependence of the mechanical response can be modeled by appropriate adjustment in dashpot and spring values, that is, [11], lower spring stiffness and dashpot viscosity levels for higher temperature and vice versa for lower temperature conditions. **Figure 7** illustrates in detail the induced strain as a function of the applied stress and loading time duration for the four models [11].

7. Equivalent mechanical models analysis

Simply, each mechanical component under direct stress, such as tensile or compressive stresses, will exhibit either elastic (temporary), Plastic (permanent), or recoverable (viscoelastic) deformation, so the following paragraphs will shed spotlight on the analogy (similarity) between behavior of the five equivalent mechanical models and an actual case of a polymeric (viscoelastic) component under direct load as illustrated below, so let us start with first one:

7.1 Maxwell model

Figure 8 shows a schematic representation for Maxwell model which contains two elements spring (E and Young's modulus) and dashpot (fluid viscosity in terms of stress- time) in series under effect of external force F, applied in two inline opposite ends, and for analysis, there are three identifying points A, B, and C.

Hence, the required parameter is the total deformation or extension δ that occurs in the whole length of this model, so

$$\delta = \delta_{AB} + \delta_{BC} \tag{22}$$

Where:

F: Applied force to a linear spring and a dashpot in series to form a Maxwell model.

E: Modulus of the spring ($K = \mathrm{EL}$).

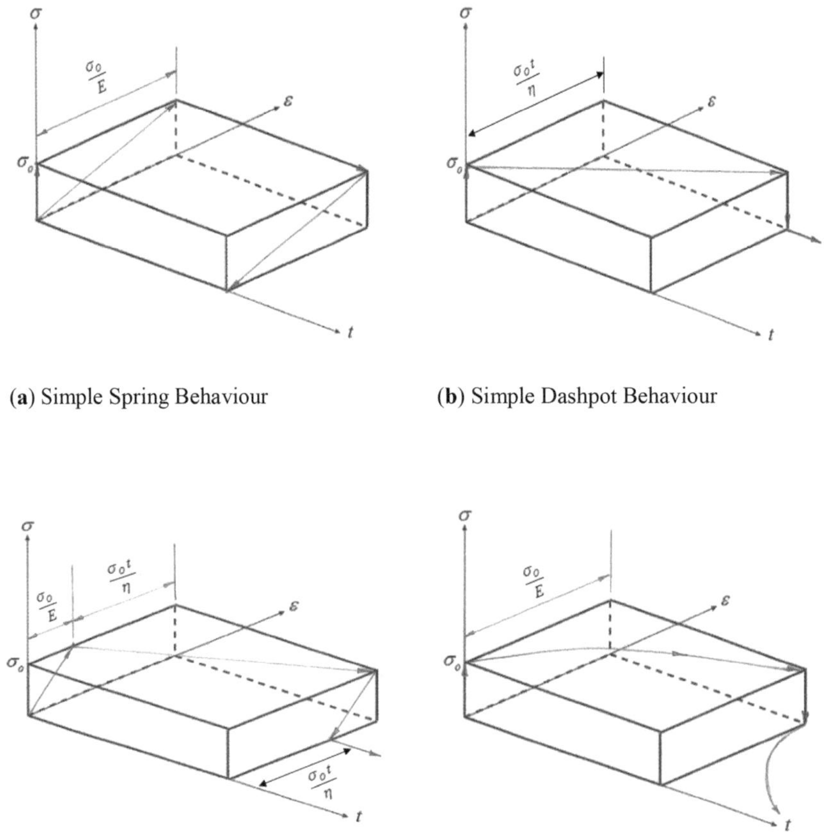

(**a**) Simple Spring Behaviour (**b**) Simple Dashpot Behaviour

(**c**) Maxwell Model Behaviour (**d**) Voigt Model Behaviour

Figure 7.
Stress-strain time diagram for mechanical analogs.

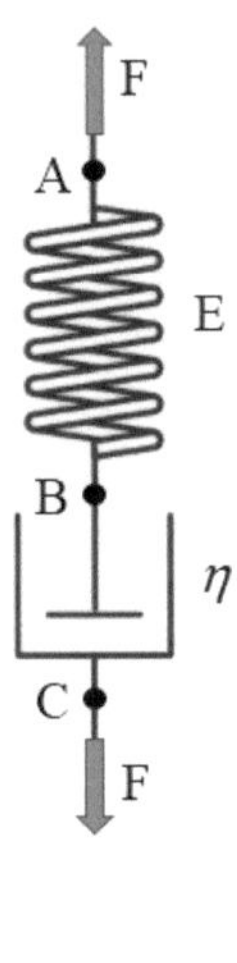

Figure 8.
Simple Maxwell model.

η: Dynamic viscosity of the dashpot
δ: Extension.

Now, by differentiating the above equation with respect to time, using D-operator method yields:

$$D\delta = D\delta_{\text{AB}} + D\delta_{\text{BC}} \tag{23}$$

For springs, Hooks law applies as follows:

$$F = K\delta_{\text{AB}} \tag{24}$$

$$F = \text{EL}\delta_{\text{AB}} \tag{25}$$

$$\text{DF} = \text{EL}\, D\delta_{\text{AB}} \tag{26}$$

$$D\delta_{\text{AB}} = \frac{\text{DF}}{\text{EL}} \tag{27}$$

For viscous element or dashpot, Newton's law of viscosity applies as follows:

$$F = \eta\, L\, D\delta_{\text{BC}} \tag{28}$$

By employing D-operator method, yields:

$$D\delta = \frac{\text{DF}}{\text{EL}} + \frac{F}{\eta L} \tag{29}$$

If (A) denotes some characteristic cross-sectional dimension, where the force F is applied and L_o is the original length of the Maxwell unit, then:

$$\sigma = \frac{F}{A} \tag{30}$$

$$\sigma = \frac{F}{L^2} \tag{31}$$

and

$$\boldsymbol{\varepsilon} = \frac{\boldsymbol{\delta}}{\boldsymbol{L_o}} \tag{32}$$

$$\varepsilon \simeq \frac{\delta}{L} \tag{33}$$

$$\sigma + \left(\frac{\eta}{E}\right)\sigma = \eta\varepsilon \tag{34}$$

$$\left[\varepsilon L = \frac{\sigma\, L^2}{\text{EL}} + \frac{\sigma L^2}{\eta L}\right] * \frac{\eta}{L} \tag{35}$$

$$F = \sigma\, L^2 \Rightarrow \text{DF} = \sigma L^2 \tag{36}$$

$$\delta = \varepsilon\, L \Rightarrow D\delta = \varepsilon L \tag{37}$$

$$\eta c = \frac{\eta}{E}\sigma \mid \sigma \tag{38}$$

Here, the dot () notation has been used in place of the D-operator. A comparison of the above first and last equations shows that the latter expression for Hook's law.

Where:

$a_o = 1, a_1 = \frac{\eta}{E}, b_o = 0, b_1 = \eta$

and the remaining constants are zeros, so

$$\left[a_o + a_1\frac{\partial}{\partial t} + \dots \dots\right]\sigma = \left[b_o + b_1\frac{\partial}{\partial t} + \dots\right]\varepsilon \tag{39}$$

$$a_o\sigma + a_1\frac{\partial\sigma}{\partial t} + \dots \dots = b_o\varepsilon + b_1\frac{\partial\varepsilon}{\partial t} + \dots \tag{40}$$

$$a_o\sigma + a_1\sigma + \dots \dots = b_o\varepsilon + b_1\varepsilon + \dots \tag{41}$$

$$1\sigma + \frac{\eta}{E}\sigma + \dots \dots = 0\varepsilon + \eta\varepsilon + \dots \tag{42}$$

now by comparing the above two equations, yields:

$$a_o = 1, a_1 = \frac{\eta}{E}, b_o = 0, b_1 = \eta$$

and the remaining constants are zeros.

7.1.1 Maxwell boundary conditions

It is noted that, for modulus of elasticity approaches infinity ($E \Rightarrow \infty$), the general equation for Maxwell model will be reduced to a simple dashpot only as indicated in the following set of equations:

$$D\delta = \frac{\mathrm{DF}}{\mathrm{EL}} + \frac{F}{\eta L} \tag{43}$$

$$D\delta = \frac{F}{\eta L} \quad \text{for a dashpot only} \tag{44}$$

Whereas, for fluid viscosity approaches infinity ($\eta \Rightarrow \infty$), in the general equation of Maxwell model [12], this specific equation will be reduced to a simple spring only exactly as illustrated in the following mathematical equations:

$$D\delta = \frac{\mathrm{DF}}{\mathrm{EL}} + \frac{F}{\eta L} \tag{45}$$

$$D\delta = \frac{\mathrm{DF}}{\mathrm{EL}} \tag{46}$$

$$\delta = \frac{F}{\mathrm{EL}} \quad \textbf{for a spring only} \tag{47}$$

7.2 Voigt model

This model has two elements, elastic element (spring) and viscous element (dashpot), connected in parallel, as shown in **Figure 9**.

Spring (E) and dashpot (η) in parallel, so

$$F = F_S + F_D \tag{48}$$

$$F_S = \mathrm{EL}\delta \tag{49}$$

$$F_D = \eta \mathrm{LD}\delta \tag{50}$$

where δ = total extension of either element of Voigt model, thus the above equation of the applied force will be in the following form:

$$F = \mathrm{EL}\delta + \eta \mathrm{LD}\delta \tag{51}$$

This equation is representing the characteristic equation of the Voigt model, and now, by analogy, it may be rewritten as follows:

$$F = \mathrm{EL}\delta + \eta \mathrm{LD}\delta \tag{52}$$

$$\sigma = E\varepsilon + \eta\varepsilon \tag{53}$$

$$\sigma = \frac{F}{L^2}, \quad \delta = \varepsilon\, L \quad \text{and} \quad D\delta = \varepsilon L \tag{54}$$

$$\frac{F}{L^2} = \frac{\mathrm{EL}}{L^2} * \varepsilon\, L + \frac{\eta L}{L^2}\varepsilon L \tag{55}$$

$$\sigma = E\varepsilon + \eta\varepsilon \tag{56}$$

Now, by comparing the obtained equation with the standard formula, yields:

$$a_o\sigma + a_1\sigma + \dots \dots = b_o\varepsilon + b_1\varepsilon + \dots \tag{57}$$

$$1\,\sigma + 0\sigma + \dots \dots = E\varepsilon + \eta\varepsilon + \dots \tag{58}$$

$$a_o = 1, a_1 = 0, b_o = E, b_1 = \eta$$

and the remaining constants are zeros.

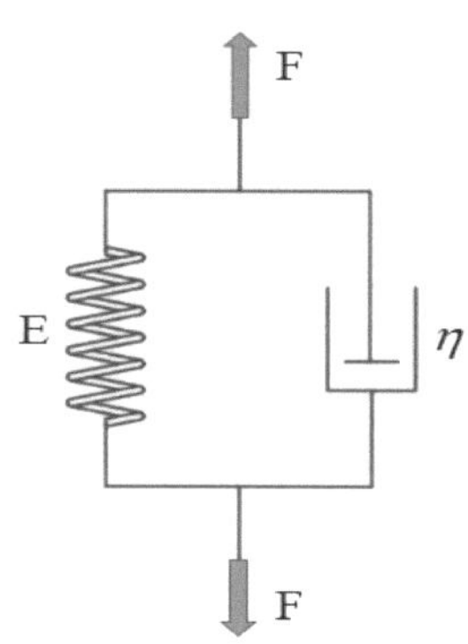

Figure 9.
Simple Voigt model.

7.2.1 Voigt boundary conditions

Similarly, in Voigt model, as the modulus of elasticity approaches zero ($E \Rightarrow 0$), the governing equation will be in the following form:

$$F = \text{EL}\delta + \eta\text{LD}\delta \tag{59}$$

$$F = \eta\text{LD}\delta \textbf{ for a dashpot only} \tag{60}$$

and on the other side if the fluid viscosity is zero ($\eta \Rightarrow 0$), then the resulting equation is:

$$F = \text{EL}\delta + \eta\text{LD}\delta \tag{61}$$

$$F = \text{EL}\delta \textbf{ for a spring only} \tag{62}$$

In these cases, the Maxwell and Voigt models are degenerating into the simplest elements. Thus, it is required to establish a nondegenerate model which is a little more complex than the two previous models considered. The coming sections will deal with a modified model for both Maxwell and Voigt.

7.3 Modified Maxwell model

This modified model is shown in **Figure 10** below and it consists of three elements; they are:

- Elastic element (spring with modulus of elasticity)
- Elastic element (spring with modulus of elasticity)
- Viscous element (dashpot with fluid viscosity)

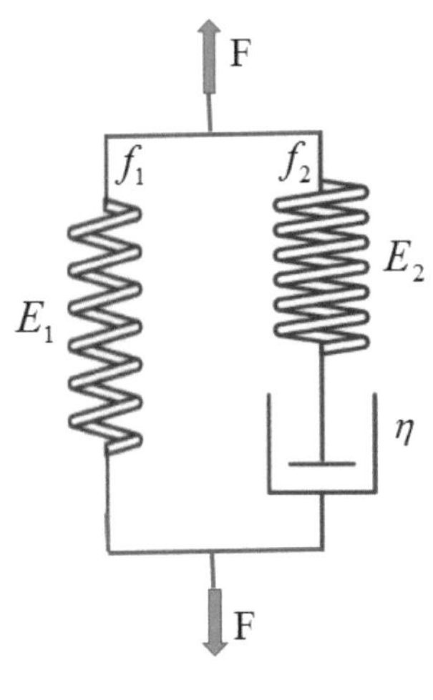

Figure 10.
Modified Maxwell model.

The applied force on the outer terminals of this model is F, but this overall force will be divided into two sub-forces f_1 and f_2.

$$F = f_1 + f_2 \tag{63}$$

$$f_1 = E_1 L\delta \text{ and } \mathrm{Df}_2 = E_1 \mathrm{LD}\delta \tag{64}$$

and in return to the general governing equation for Maxwell model, which is exactly similar to the right-hand side of the Maxwell modified model, as shown in **Figure 11.**

$$D\delta = \frac{\mathrm{DF}_2}{E_2 L} + \frac{F_2}{\eta L} \tag{65}$$

Now, it is better to eliminate both of f_1 and f_2 in order to get an analysis of the induced stresses and the associated strains as follows:

$$f_1 = E_1 L\delta \tag{66}$$

$$\mathrm{Df}_1 = E_1 \mathrm{LD}\delta \tag{67}$$

$$\mathrm{Df}_2 = E_2 \mathrm{LD}\delta - \left(\frac{E_2}{\eta}\right) f_2 \tag{68}$$

$$F = f_1 + f_2 \tag{69}$$

$$\mathrm{DF} = \mathrm{Df}_1 + \mathrm{Df}_2 \tag{70}$$

$$\mathrm{DF} = E_1 \mathrm{LD}\delta + E_2 \mathrm{LD}\delta - \left(\frac{E_2}{\eta}\right) f_2 \tag{71}$$

$$\mathrm{DF} = E_1 \mathrm{LD}\delta + E_2 \mathrm{LD}\delta - \left(\frac{E_2}{\eta}\right)(F - E_1 L\delta) \tag{72}$$

$$\mathrm{DF} = (E_1 + E_2)\mathrm{LD}\delta - \left(\frac{E_2}{\eta}\right) F + \frac{E_1 E_2 L}{\eta}\delta \tag{73}$$

$$\mathrm{DF} + \left(\frac{E_2}{\eta}\right) F = (E_1 + E_2)\mathrm{LD}\delta + \frac{E_1 E_2 L}{\eta}\delta \tag{74}$$

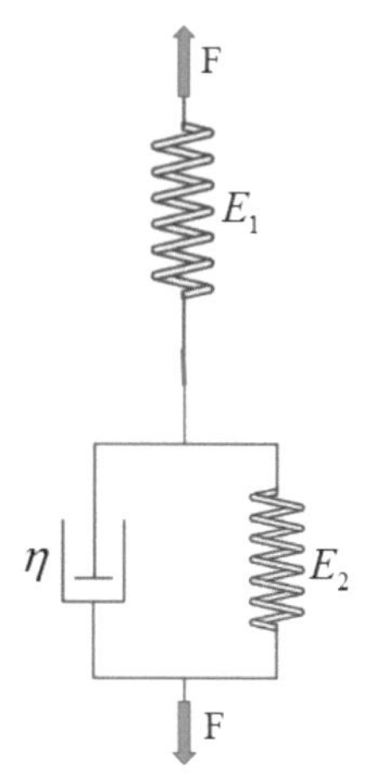

Figure 11.
Modified Voigt model.

$$\left[\mathrm{DF} + \left(\frac{E_2}{\eta}\right)F = (E_1 + E_2)L^2\varepsilon + \frac{E_1E_2L^2}{\eta}\varepsilon\right]\frac{\eta}{L^2E_2} \tag{75}$$

$$\sigma = \frac{F}{L^2} \quad \Rightarrow \sigma = \frac{\mathrm{DF}}{L^2} \tag{76}$$

$$\varepsilon = \frac{\delta}{L} \Rightarrow \quad \delta = \varepsilon\, L \quad \Rightarrow D\delta = \varepsilon\, L \tag{77}$$

$$f_2 = F - f_1 \tag{78}$$

$$f_2 = F - E_1L\delta \tag{79}$$

$$\left(\frac{\eta}{E_2}\right)\sigma + \sigma = \left(\frac{E_1 + E_2}{E_2}\right)\eta\varepsilon + E_1\varepsilon \tag{80}$$

$$\sigma + \left(\frac{\eta}{E_2}\right)\sigma = E_1\varepsilon + \left(\frac{E_1 + E_2}{E_2}\right)\eta\varepsilon \tag{81}$$

$$a_o\sigma + a_1\sigma + \dots\dots = b_o\varepsilon + b_1\varepsilon + \dots \tag{82}$$

$$1\,\sigma + \left(\frac{\eta}{E_2}\right)\sigma = E_1\varepsilon + \left(\frac{E_1 + E_2}{E_2}\right)\eta\varepsilon \tag{83}$$

By comparing the last two main equations, yields:

$$a_o = 1, a_1 = \left(\frac{\eta}{E_2}\right), b_o = E_1, b_1 = \left(\frac{E_1 + E_2}{E_2}\right)\eta \tag{84}$$

and the other constants and terms with higher orders are zeros.

The modified Maxwell model has the advantage over time period and is avoided in the separate spring element E_1, whereas it is permitted to occur in the left-hand branch of the sketch in **Figure 10**. Such limited relaxation behavior is typical of polymers and elastomers subjected to a long-duration stress environment, so that the modified Maxwell model (although still highly simplistic in terms of actual material performance) depicts the nature of viscoelastic behavior in a much more realistic manner [13].

7.4 Modified Voigt model

The full detail sketch for the modified Voigt model is shown in **Figure 11.** Where this model is consisting of the following elements:

- Elastic element (spring with modulus of elasticity E_1)
- Elastic element (spring with modulus of elasticity E_2)
- Viscous element (dashpot with fluid viscosity η).

And the external applied force is denoted by F.

Based on the given layout, the total deformation along the external terminals is δ and may be expressed in terms of the sub-deformation components as shown in the coming mathematical equation:

$$\delta = \delta_1 + \delta_2 \tag{85}$$

also, for the spring element E_1, there is:

$$F = E_1 L \delta_1 \tag{86}$$

and for simple Voigt model:

$$F = E_2 L \delta_2 + \eta \mathrm{LD}\delta_2 \tag{87}$$

Now, by eliminating δ_1 and δ_2 from the above equation, yields:

$$\delta = \delta_1 + \delta_2 \tag{88}$$

$$D\delta = D\delta_1 + D\delta_2 \tag{89}$$

$$\delta_2 = \delta + \delta_2 \tag{90}$$

$$\delta_2 = \delta + \frac{F}{E_1 L} \tag{91}$$

$$F = E_1 L \delta_1 \Rightarrow \mathrm{DF} = E_1 \mathrm{LD}\delta_1 \Rightarrow D\delta_1 = \frac{\mathrm{DF}}{E_1 L} \tag{92}$$

$$F = E_2 L \delta_2 + \eta\ \mathrm{LD}\delta_2 \tag{93}$$

$$\eta\ \mathrm{LD}\delta_2 = F - E_2 L \delta_2 \Rightarrow\ D\delta_2 = \frac{F}{\eta L} - \frac{E_2}{\eta}\delta_2 \tag{94}$$

$$D\delta = \frac{\mathrm{DF}}{E_1 L} + \frac{F}{\eta\, L} - \frac{E_2}{\eta}\delta_2 \tag{95}$$

$$D\delta = \frac{\mathrm{DF}}{E_1 L} + \frac{F}{\eta\, L} - \frac{E_2}{\eta}\left(\delta - \frac{F}{E_1 L}\right) \tag{96}$$

$$\left[D\delta = \frac{\mathrm{DF}}{E_1 L} + \frac{F}{\eta\, L} - \frac{E_2}{\eta}\delta + \frac{E_2}{E_1 \eta L}F\right] * E_1 L \tag{97}$$

$$E_1 \mathrm{LD}\delta = \mathrm{DF} + \frac{E_1}{\eta}F - \frac{E_1 E_2 L}{\eta}\delta + \frac{E_2}{\eta}F \tag{98}$$

$$E_1 \mathrm{LD}\delta = \mathrm{DF} + \left(\frac{E_1 + E_2}{\eta}\right)F - \left(\frac{E_1 E_2 L}{\eta}\right)\delta \tag{99}$$

Now, by using separation of variables, yields:

$$\mathrm{DF} + \left(\frac{E_1 + E_2}{\eta}\right)F = E_1 \mathrm{LD}\delta + \left(\frac{E_1 E_2 L}{\eta}\right)\delta \tag{100}$$

$$\delta = \varepsilon\, L \qquad \Rightarrow D\delta = \varepsilon\, L \tag{101}$$

$$\mathrm{DF} + \left(\frac{E_1 + E_2}{\eta}\right)F = E_1 L^2 \varepsilon + \left(\frac{E_1 E_2 L^2}{\eta}\right)\varepsilon \tag{102}$$

$$\left[\mathrm{DF} + \left(\frac{E_1 + E_2}{\eta}\right)F = E_1 L^2 \varepsilon + \left(\frac{E_1 E_2 L^2}{\eta}\right)\varepsilon\right] * \frac{\eta}{(E_1 + E_2)L^2} \tag{103}$$

$$\frac{F}{L^2} + \left(\frac{\eta}{E_1 + E_2}\right)\frac{\mathrm{DF}}{L^2} = \left(\frac{E_1 E_2}{E_1 + E_2}\right)\varepsilon + \left(\frac{E_1 \eta}{E_1 + E_2}\right)\varepsilon \tag{104}$$

$$\sigma + \left(\frac{\eta}{E_1 + E_2}\right)\sigma = \left(\frac{E_1E_2}{E_1 + E_2}\right)\varepsilon + \left(\frac{E_1\eta}{E_1 + E_2}\right)\varepsilon \tag{105}$$

$$a_o\sigma + a_1\sigma + \dots \dots = b_o\varepsilon + b_1\varepsilon + \dots \tag{106}$$

Now, by comparing the last two equations, yields:

$$a_o = 1, a_1 = \left(\frac{\eta}{E_1 + E_2}\right), b_o = \left(\frac{E_1E_2}{E_1 + E_2}\right), b_1 = \left(\frac{E_1\,\eta}{E_1 + E_2}\right) \tag{107}$$

And the other higher order terms are zeros.
In return to these two equations,
The first one is from the modified Maxwell model analysis:

$$\mathrm{DF} + \left(\frac{E_2}{\eta}\right)F = (E_1 + E_2)\mathrm{LD}\delta + \frac{E_1E_2L}{\eta}\delta \tag{108}$$

and the second one is from the modified Voigt model analysis:

$$\mathrm{DF} + \left(\frac{\mathrm{E}_1 + \mathrm{E}_2}{\eta}\right)\mathrm{F} = \mathrm{E}_1\mathrm{LD}\delta + \left(\frac{\mathrm{E}_1\mathrm{E}_2\mathrm{L}}{\eta}\right)\delta \tag{109}$$

These two last equations are having essentially the same form, although the constants vary depending on which model is finally selected, so the general form of these equations is:

$$\mathrm{DF} + P_oF = q_1\,D\delta + q_o\delta \tag{110}$$

These representations of viscoelastic behavior by either the modified Maxwell or modified Voigt model are identical. Both modified models are having two springs and one viscous element. Alternatively, it is possible to use one spring and two viscous elements, which would yield the following form of force deflection equation as shown in the following equation:

$$\mathrm{DF} + P_oF = q_2\,D^2\delta + q_1D\delta \tag{111}$$

7.5 Maxwell-Voigt model

This specific model consists of Maxwell model connected in series with Voigt model as illustrated in **Figure 12.** Where the main components are:
Elastic element (spring with modulus of elasticity E_1)

- Elastic element (spring with modulus of elasticity E_2)
- Viscous element (dashpot with fluid viscosity η_1)
- Viscous element (dashpot with fluid viscosity η_2)

Let δ_1 and δ_2 represent the induced deformation in the Maxwell and Voigt elements, respectively, as a result of applying the external force F to both ends of the compound model [14], **Figure 12**, then:

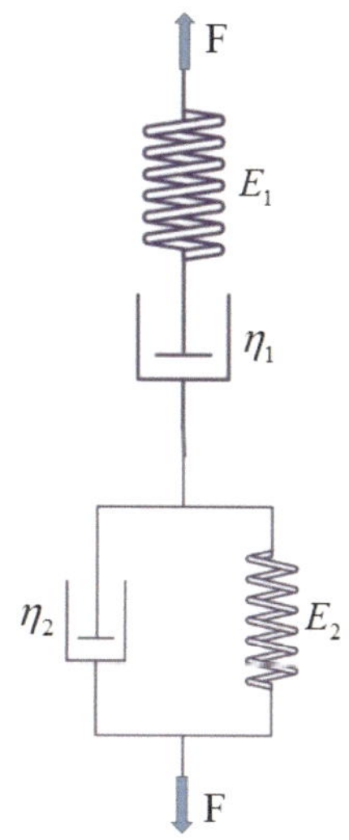

Figure 12.
Maxwell-Voigt model.

$$\delta = \delta_1 + \delta_2 \tag{112}$$

where the δ is the total elongation with length dimension. It is required to analyze this compound model by considering each model alone, as indicated in **Figure 13**.

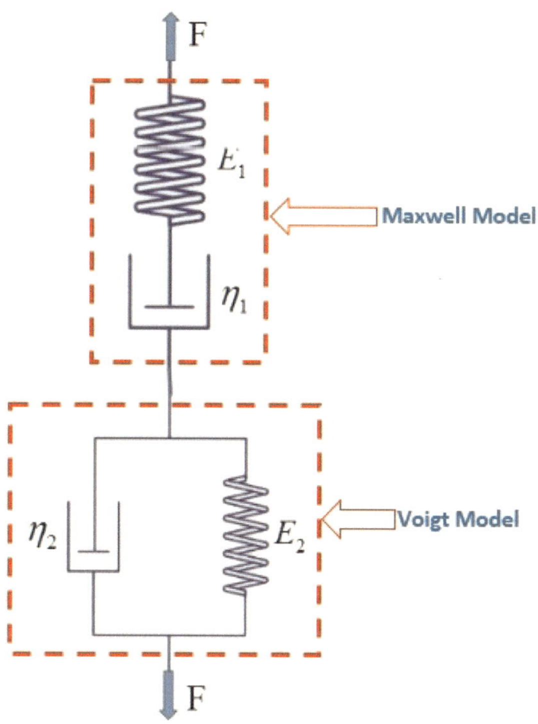

Figure 13.
Maxwell model in series with Voigt model.

Now, for Maxwell element, it is already stated that:

$$D\delta_1 = \frac{DF}{E_1 L} + \frac{F}{\eta_1 L} \tag{113}$$

and for Voigt model:

$$F = F_S + F_D \tag{114}$$

which also may be written as:

$$F = E_2 L\delta_2 + \eta_2 LD\delta_2 \tag{115}$$

It is possible to eliminate δ_1 and δ_2 that the final result will be in the following form:

$$D^2F + \left[\frac{E_1}{\eta_1} + \frac{E_1}{\eta_2} + \frac{E_2}{\eta_2}\right] DF + \left[\frac{E_1E_2}{\eta_1\eta_2}\right] F = E_1 LD^2\delta + \left[\frac{E_1E_2L}{\eta_2}\right] D\delta \tag{116}$$

This essential equation may be rewritten in terms of stress and strain as follows:

$$\sigma + \left[\frac{E_1}{\eta_1} + \frac{E_1}{\eta_2} + \frac{E_2}{\eta_2}\right]\left(\frac{\eta_1\eta_2}{E_1E_2}\right)\sigma + \left(\frac{\eta_1\eta_2}{E_1E_2}\right)\overset{\dot{=}\eta_1\varepsilon + \left(\frac{\eta_1\eta_2}{E_2}\right)\dot{\varepsilon}}{\sigma} \tag{117}$$

and according to the general formula in the following equation:

$$a_o\sigma + a_1\sigma + a_2 \overset{\therefore\ \ldots = b_o\varepsilon + b_1\varepsilon + b_2 \overset{+}{\varepsilon}\cdots}{\sigma} \tag{118}$$

$$a_o = 1, a_1 = \left(\frac{E_1}{\eta_1} + \frac{E_1}{\eta_2} + \frac{E_2}{\eta_2}\right)\left(\frac{\eta_1\eta_2}{E_1E_2}\right), a_2 = \left(\frac{\eta_1\eta_2}{E_1E_2}\right) \tag{119}$$

$$b_o = 0, b_1 = \eta_1, b_2 = \left(\frac{\eta_1\eta_2}{E_2}\right) \tag{120}$$

The relationships derived in the previous mechanical models are specific forms of the generalized Hook's law relating stress to strain, in accordance with the following equation [15]:

$$\left[a_o + a_1\frac{\partial}{\partial t} + a_2\frac{\partial^2}{\partial t^2} + \ldots\ldots\right]\sigma = \left[b_o + b_1\frac{\partial}{\partial t} + b_2\frac{\partial^2}{\partial t^2} + \ldots\ldots\right]\varepsilon \tag{121}$$

However, there has been no limitation on the time history of the applied force F or the induced deformation δ in other words these equations are representing the main basis of behavior of a viscoelastic material suffering from creep and stress relaxation. So to analyze any viscoelastic materials, means studying the relationship between the applied know force and the associated viscoelastic strain, it is so essential to know the numerical values of the included mechanical components (elastic element, spring and viscous element, and dashpot), so **Table 2** summarize these constants in detail based on the type of the proposed equivalent model.

By applying a constant force to previously unloaded models, the extension when measured as a function of time over a long period, is called creep motion or creep [16].

b_2	b_1	b_o	a_2	a_1	a_o	Schematic	Model
0	0	0	0	$\frac{\eta}{E}$	1	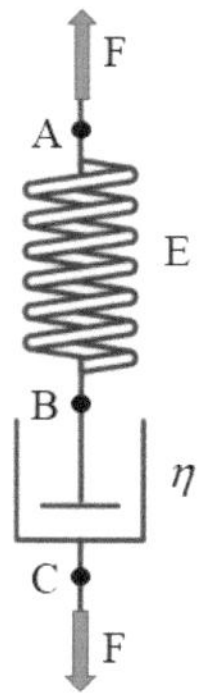	Maxwell
0	η	E	0	0	1	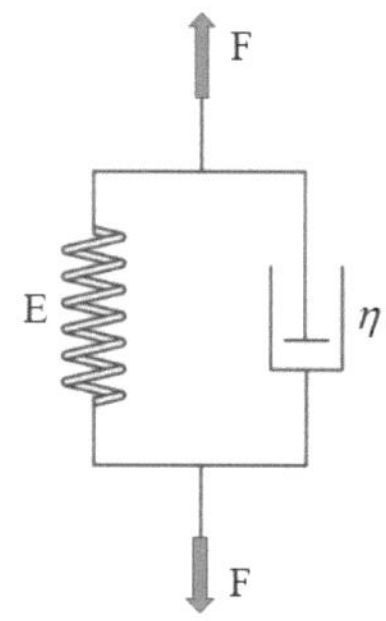	Voigt
0	$\left(\frac{E_1+E_2}{E_2}\right)\eta$	E_1	0	$\frac{\eta}{E_2}$	1	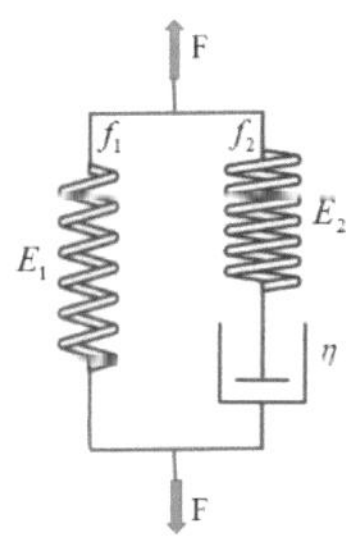	Modified Maxwell
0	$\left(\frac{E_1}{E_1+E_2}\right)\eta$	$\left(\frac{E_1E_2}{E_1+E_2}\right)$	0	$\frac{\eta}{E_1+E_2}$	1	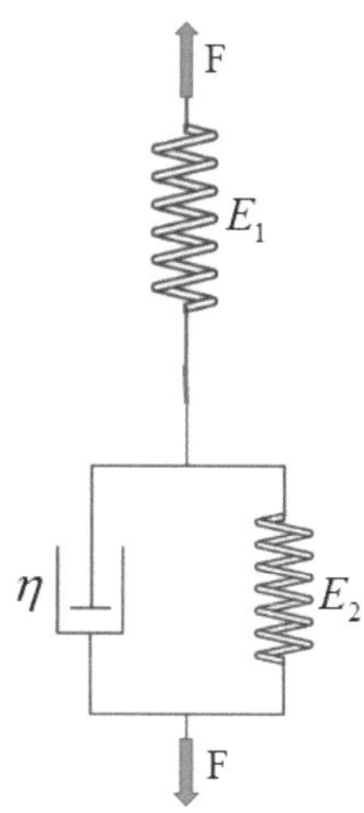	Modified Voigt

b_2	b_1	b_o	a_2	a_1	a_o	Schematic	Model
$\frac{\eta_1\eta_2}{E_2}$	η_1	0	$\left(\frac{\eta_1\eta_2}{E_1E_2}\right)$	$\left(\frac{E_1}{\eta_1}+\frac{E_1}{\eta_2}+\frac{E_2}{\eta_2}\right)\left(\frac{\eta_1\eta_2}{E_1E_2}\right)$	1	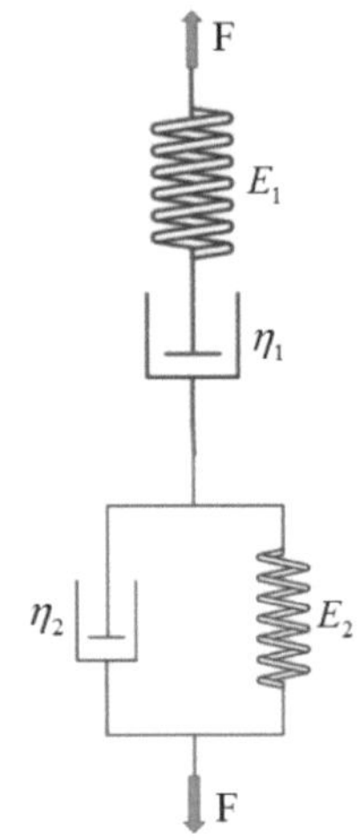	Maxwell Voigt

Table 2.
Values of the constants in the generalized Hook's law for the proposed mechanical models.

Conversely, the application of a constant extension to previously unloaded models require a time-dependent force that can be measured. This force decreases in time accordingly, as a relaxation effect occurs within the model.

7.6 Creep response

Given a force, F, according to the relationship:

$$F = \mathrm{CH}(t) \tag{122}$$

where C is an arbitrary constant, and $H(t)$ is the **Heaviside function** of time and has the following values:

$$H(t) = 0 \text{ for } t \leq 0$$
$$H(t) = 1 \text{ for } t \geq 1$$

The time derivative of the function $H(t)$ is the well-known **Dirac function** (Δ), thus:

$$\Delta(t) = \mathrm{DH}(t) \tag{123}$$

Both $H(t)$ and $\Delta(t)$ are shown as a function of time, see **Figure 14**.

So, it is required to find the resulting extension $\delta(t)$ for the entire model, for a simple spring element with spring modulus E:

$$\delta(t) = \frac{F}{\mathrm{EL}} = \frac{C}{\mathrm{EL}} H(t) \tag{124}$$

And for the viscous element with viscosity η:

$$D\delta(t) = \frac{C}{\eta L} H(t) \tag{125}$$

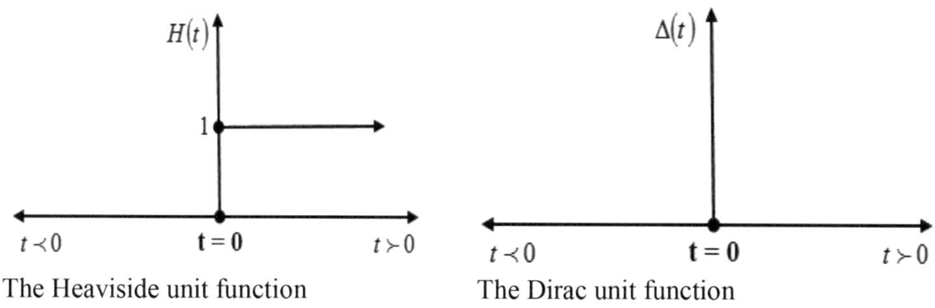

Figure 14.
Heaviside and Dirac unit functions. The Heaviside unit function. The Dirac unit function.

This differential equation can be integrated to give:

$$\delta(t) = \frac{C}{\eta L} t H(t) \tag{126}$$

The constant of integration in the last equation is zero since $\delta(t) = 0$ at $t = 0$. For Maxwell element, the application of the above last equations gives:

$$D\delta(t) = \frac{\text{DF}}{\text{EL}} + \frac{F}{\eta L} \tag{127}$$

$$D\delta(t) = \frac{C}{\text{EL}}\Delta(t) + \frac{C}{\eta L}H(t) \tag{128}$$

which after integration shows the following time-dependent response:

$$\delta(t) = \frac{C}{\text{EL}}H(t) + \frac{C t}{\eta L}H(t) \tag{129}$$

$$\delta(t) = \frac{C}{L}\left(\frac{1}{E} + \frac{t}{\eta}\right) H(t) \tag{130}$$

This last equation is representing the overall extension (deformation) for Maxwell model.

For Voigt model the governing equation is as follows:

$$F = \text{EL}\delta + \eta\text{LD}\delta \tag{131}$$

$$\text{CH}(t) = \text{EL}\delta(t) + \eta\text{LD}\delta(t) \tag{132}$$

In order to solve the above differential equation, it is so essential to multiply each side of this specific equation by an integrating factor:

Integrating Factor $= e^{\left(\frac{E}{\eta}\right)t}$, so that:

$$\text{C}e^{\left(\frac{E}{\eta}\right)t}H(t) = \text{EL}\delta(t)e^{\left(\frac{E}{\eta}\right)t} + \eta\text{LD}\delta(t)e^{\left(\frac{E}{\eta}\right)t} \tag{133}$$

$$\text{C}e^{\left(\frac{E}{\eta}\right)t}H(t) = \eta\text{LD}\left[\delta(t)e^{\left(\frac{E}{\eta}\right)t}\right] \tag{134}$$

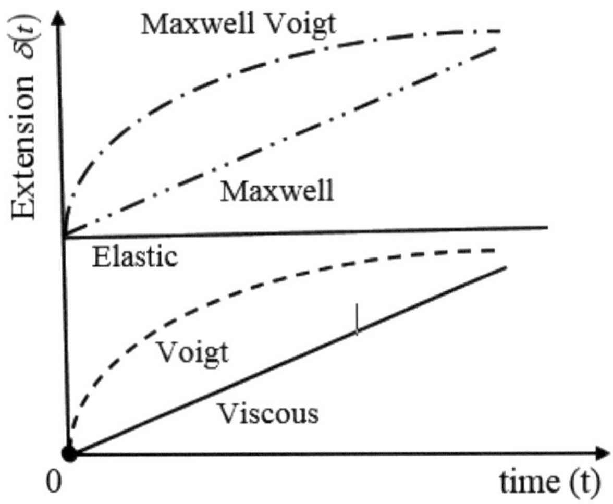

Figure 15.
Extension response (creep behavior) of the proposed mechanical models.

after integration yields:

$$\delta(t) = \frac{C}{\mathrm{EL}}\left[1 - e^{-\left(\frac{E}{\eta}\right)t}\right] H(t) \tag{135}$$

This equation is representing the induced response of Voigt model. The extension response (creep behavior) of mechanical models, including the Maxwell-Voigt, modified Maxwell, modified Voigt, and simple elements models [17], is shown in **Figure 15**.

It is to be noted that the Maxwell element has a response equal to sum of the responses for the viscous and elastic elements, because it consists of these elements in series. Also, the Maxwell Voigt model consists of Maxwell and Voigt models in series, so that its response is given the following equation:

$$\delta(t) = C\left[\frac{1}{E_1 L} + \frac{t}{\eta L}\right] H(t) + \frac{C}{E_2 L}\left[1 - e^{-\left(\frac{E}{\eta}\right)}t\right] H(t) \tag{136}$$

8. Discussion

In most cases of nominating a material for manufacturing an industrial component, and sustaining the applied load, usually either tensile, compressive, or shear stress, it is so essential to seek the best mechanical properties, including Young's modulus (modulus of elasticity) and the associated strain(s). So, this theoretical analysis via creating a mechanical model for describing a polymeric material under stress and how such specific material will behave or exhibit a resistance during loading phase and what is the expected result—deformation style as a function of both loading time duration and the ambient temperature, so the first assumption was depending on a simple linear spring (elastic element) but after removing the applied load the spring will utterly return to its original dimensions that if the applied load within the elastic limit of this spring. Hence, this spring will not well cover the actual behavior of the loaded polymeric component, so the next proposal was considering only viscous element —dashpot but unfortunately this element also was not able to describe the gained behavior, these two results shed light on the combined models, so the next models was Maxwell, Voigt, modified Maxwell, modified Voigt, and Maxwell-Voigt models, the above theoretical equations supported by the finding shown in **Figure 15** gives a graduate interpretation of the expected behavior of the polymeric material under direct load and under constant room temperature, but it is very accurate to say that Maxwell-Voigt model is well telling or drawing the gained path of the residual strain in the polymeric material, so this is invitation to extend ideas for more models and more analysis for reaching the best behavior analogy between the mechanical models and polymeric materials as well.

9. Conclusions

As a result of this theoretical study, it is essential to conclude that all of the analyzed models are giving an interpretation of the behavior of some loaded polymeric materials but with different approximation but it looks like that each model is complementing the other models, or in other words, starting from simple spring alone or viscous dashpot alone is giving high error rate and this error starts decreases via adding another element either in series or in parallel to form Maxwell and Voigt models to reach optimum verification, respectively, so the modified Maxwell model is giving relatively the best fit with the actual behavior of the polymeric component under direct longitudinal load, but that is not meaning neglecting the other models results, especially both of Voigt and Maxwell-Voigt models. It is preferable to include a conclusion(s) section, which will summarize the content of the book chapter.

Acknowledgements

We wish to acknowledge the help provided by the technical and support staff in the Mechanical Power Engineering Department, Mussaib Technical College, Al Furat Al Awsat Technical University. We would also like to show our deep appreciation to Prof. Dr. Muhsin J. Jweeg who helped us finalize our book chapter.

Nomenclature

σ	Stress
σ_o	initial applied stress
$\sigma(t)$	stress as a function of time (t)
σ	first derivative of stress with respect to time
$\dot{\sigma}$	second derivative of stress with respect to time
ε	Strain
ε_o	induced strain
$\varepsilon(t)$	strain as a function of time (t)
ε	first derivative of strain with respect to time
$\dot{\varepsilon}$	second derivative of strain with respect to time
$\varepsilon_{critical}$	critical strain
γ	shear strain
γ	shear-strain rate

δ	deformation (longitudinal extension)
$\delta(t)$	deformation as a function of time
η	fluid viscosity of the employed dashpot
E	Young's modulus (modulus of elasticity)
G	shear modulus
$E_c(t)$	creep modulus as a function of time
$E_r(t)$	relaxation modulus as a function of time
F	applied external force
t	time
t	relaxation time
T	Temperature
K	spring stiffness
C	damping constant
ΔH	viscous flow activation energy at a particular temperature
R	universal gas constant
A	pre-exponent factor

Author details

Emad Kamil Hussein[1*], Batool Mardan Faisal[2], Kussay Ahmed Subhi[3], Thiago Santos[4], Samir Ghouali[5], Muhammad Asyraf[6] and Caroliny Santos[4]

1 Mechanical Power Engineering Department, Mussaib Technical College, Al Furat Al Awsat Technical University, Babil, Iraq

2 Mechanical Engineering Department, College of Engineering, Wasit University, Wasit, Iraq

3 Mechanical Equipment and Machines Department, Mussaib Technical College, Al Furat Al Awsat Technical University, Babil, Iraq

4 Textiles Technologies Study Group (GETTEX), Laboratory of Knitting, Federal University of Rio Grande do Norte, Natal, Rio Grande do Norte, Brazil

5 Mascara, Algeria & STIC Laboratory, Faculty of Sciences and Technology, Mustapha Stambouli University, University of Tlemcen, Algeria

6 Engineering Design Research Group (EDRG), Faculty of Engineering, School of Mechanical Engineering, Universiti Teknologi Malaysia (UTM), Johor, Malaysia

*Address all correspondence to: emad_kamil72@atu.edu.iq

References

[1] Shenoy AV, Saini DR. Thermoplastic Melt Rheology and Processing. Madison Avenue, New York; 1996. p. 10016

[2] Marques P, Creus J. Computational Viscoelasticity. Springer Science & Business Media; 2012

[3] Dulgerova P. Biophysical modelling of viscoelastic deformations of cattle skin in vivo. Trakia Journal of Sciences. 2004;**2**(4):1-3

[4] Faith A. Understanding Rheology. Morrison: Oxford University Press; 2001

[5] Hussein E, A Modified Version of the Prosthetic Foot [thesis]. Nahrain University; 2011

[6] Meral F. Fractional calculus in viscoelasticity: An experimental study. Communication Nonlinear Science Numerical Simulation. 2010;**15**:939. DOI: 10.1016/j.cnsns.2009.05.004

[7] Korcz H, Polybutylene polymers for hot melt adhesives, Adhesive Age, 1984; 19-23.

[8] Brinson F, & Brinson L. Polymer engineering science and viscoelasticity. An introduction, 2008;99-157.

[9] Hussein EK. Theoretical and Applied Rheology [Internet], 2017. Available from: https://www.researchgate.net/publication/307882579

[10] Adams M, Briscoe B, Johnson S. Friction and lubrication of human skin. Tribology Letters. 2007;**26**(3):239-253

[11] Soldatenkov IA. Calculation of the Deformation Component of the Force of Friction for a Standard Elastoviscous Base. Journal of Friction and Wear. 2008;**29**(1):7-14

[12] Moore DF. The Friction and Lubrication of Elastomers. Oxford: Pergamon Press; 1972

[13] Masen MA. A systems based experimental approach to tactile friction. Journal of the Mechanical Behavior of Biomedical Materials. 2011;**4**:1620-1626

[14] Sinha SK, Chong M, Lim SC. Scratching of polymers – Modelling abrasive wear. Wear. 2007;**262**: 1038-1047

[15] Kussay AS, Tudor A, Hussein EK, Wahad HS. The adhesion and hysteresis effect in friction skin with artificial materials. In: 13th International Conference on Tribology, ROTRIB'16, IOP Conference Series: Materials Science and Engineering. Vol. 174. 2017. p. 012018

[16] Hussein EK, Subhi KA, Gaaz TS. Effect of Stick-Slip Phenomena between Human Skin and UHMW Polyethylene. Pertanika. Journal of Science & Technology. 2021;**29**(3)

[17] Cua AB, Wilhelm KP, Maibach HI. Elastic properties of human skin: relation to age, sex, and anatomical region. Archives of Dermatological Research. 1990;**282**(5):283-288